MOON

AN ILLUSTRATED HISTORY

From ANCIENT MYTHS to
the COLONIES OF TOMORROW

MOON

AN ILLUSTRATED HISTORY

DAVID WARMFLASH

STERLING
New York

STERLING
New York

An Imprint of Sterling Publishing Co., Inc.
1166 Avenue of the Americas
New York. NY 10036

ISBN 978-1-4549-3198-0

Distributed in Canada by Sterling Publishing Co., Inc.
c/o Canadian Manda Group, 664 Annette Street
Toronto, Ontario M6S 2C8, Canada
Distributed in the United Kingdom by GMC Distribution Services
Castle Place, 166 High Street, Lewes, East Sussex BN7 1XU, England
Distributed in Australia by NewSouth Books
University of New South Wales, Sydney, NSW 2052, Australia

For information about custom editions, special sales,
and premium and corporate purchases,
please contact Sterling Special Sales at 800-805-5489 or
specialsales@sterlingpublishing.com.

Manufactured in the United States of America

2 4 6 8 10 9 7 5 3 1

sterlingpublishing.com

Interior design by Christine Heun and Barbara Balch
Cover design by Elizabeth Mihaltse Lindy
For Image Credits, see page 211.

To Evi, Liat, and Lior

CONTENTS

INTRODUCTION

F OR MANY OF US, THE MISSIONS OF NASA'S Project Apollo are synonymous with lunar exploration. Yet Apollo is only part of a history of the Moon that weaves all through human history—and still further back in time. In this book, we begin our exploration a little more than 4.5 billion years ago, with the formation of the Moon itself, and then work our way forward, focusing on a hundred "moments" in the history of the Moon. In selecting moments to describe, I considered current knowledge, theories, hypotheses, and thinking in lunar science; factors that had a major impact on human civilization, such as the Moon's role in calendars, ancient religions, and the emergence of astronomy and science; and, of course, the exploration of the Moon in the mid-twentieth century and the several decades of technological development leading up to that exploration. In researching the topics, I came across some interesting cultural tidbits, such as the 1929 German film Frau im Mond (Woman in the Moon), featuring a technical adviser who was key in the development of rocket technology.

As you read through this book, you'll notice that many of the moments fall into groups. The early moments describe geological events. Later, a large collection covers the influence of the Moon on humans of ancient civilizations, particularly in Mesopotamia and the Greek world. Next comes a series of related moments set during the Middle Ages, when the focal point of lunar astronomy, astronomy in general, and really the pursuit of knowledge lay within Arabic civilization, encompassing various nations and individuals that had in common the fact that they used Arabic as the language of intellectual activity. We then move through the European Renaissance to early modern times, when the advent of the telescope revealed details of land features on the Moon, as well as other phenomena, such as the fact that the planet Venus goes through phases just like the Moon. First seen by Galileo Galilei, the phases of Venus constituted one of several turning points in an ongoing debate as to whether the Sun and planets orbited Earth, or Earth orbited the Sun, a debate that emerged in ancient Greece, with various discoveries in which the Moon played a supporting role.

Entries involving the formation of lunar craters and other land features billions of years ago also introduce several heroes from human history whose work is related to the Moon and for whom craters have been named. One example, coming into our story so many times that we could call him a protagonist of the early portion of the book, is Aristarchus of Samos. Eponym of the Aristarchus crater, which is roughly 450 million years old, Aristarchus used a very creative, but simple, method to figure out the Moon's size and distance from Earth, and the distance and size of the Sun. Computing that the Sun was much bigger than the Earth, he then deduced that Earth must orbit the Sun, not the other way around. Known as the heliocentric (Sun-centered) model, the cosmology of Aristarchus won only a handful of followers in ancient times, and for many centuries a geocentric (Earth-centered) model dominated. Partly related to astronomical measurements of the Moon's motion, hints that there were problems with the geocentric model popped up during the Arabic period, leading ultimately to a revival of the heliocentric model in the work of Nicolaus Copernicus, Galileo, and Johannes Kepler, all of whom have lunar craters named for them.

Blocks of entries become increasingly linked together in the twentieth century as we look at the emergence of rocket propulsion, which ultimately would lead to NASA's Saturn V booster and the Space Race between the United States and Soviet Union. The resulting underlying story thus brings forth additional personalities, one being Sergey Korolyov, architect of the early Soviet space program. US president John F. Kennedy emerges as a protagonist in the race to the Moon because of his influence during his short time in office, whereas President Richard Nixon may strike you as more of an antagonist. You'll see why.

As noted above, there is a strong emphasis not just on events that happened prior to Apollo 11 and the first human landing on another world,

The unmanned Apollo 4 launched on November 9, 1967. It was the first flight of the Saturn V rocket, which eventually took astronauts to the Moon on the Apollo 8 mission.

but also on the missions that followed it—particularly the science that the astronauts and the Earth-based researchers carried out. Much of what we know about the Moon, and a lot of what we know about the early history of the Earth and other inner planets, comes from the studies carried out on the Moon during Project Apollo, and from the study of the 382 kilograms (842 pounds) of lunar material that those astronauts returned to Earth. The twelve astronauts who collected those samples knew what they were doing, having received hundreds of hours of training in field geology—enough training to equal a master's degree in geology, except in the case of one lunar astronaut, Harrison Schmidt, who held a PhD in geology when he arrived on the Moon during Apollo 17.

Regarding units of measurement, particularly in the Apollo-era moments, you'll notice that I employ metric, British Imperial, and nautical measurements. I did this to keep with a practice of Project Apollo: engineering values, pressure, and thrust for example, were expressed in Imperial units; distances, such as travel distance to the Moon, altitudes of orbits, and so forth were expressed in nautical miles; and measurements related to science (quantities of lunar material collected, for instance) were expressed using the metric system. (Earlier in the book, I also mention another type of unit, the stadion, an ancient Greek unit of distance, corresponding to the length of an Olympic stadium.)

The final phase of this book contemplates the near future, focusing on topics such as industrialization, solar power from the Moon,

MOON FACTS

Mass: 7.35×10^{22} kg (approximately 1/100th of Earth's mass)

Density: 3.34 g/cm^3

Surface Gravity: 1.62 m/s^2 (approximately 16 percent that on Earth's surface)

Diameter: 3,475 km (2,159 mi)

Average Orbital Distance: 384,400 km (207,343 nautical miles, 238,607 statute miles)

Perigee (minimum distance from Earth): 362,600 km

Apogee (maximum distance from Earth): 405,400 km

Sidereal Month (average time for the Moon to orbit 360° around Earth relative to the "fixed" stars): 27.32 days

Synodic Month (average time for the Moon to make one orbit around Earth with respect to the line joining the Earth and Sun): 29.53 days

Surface Temperature: –233°C to 123°C (–387° F to 253° F)

and the potential of building a large human base by the 2040s. These and other developments are on the horizon. This book's publication coincides with the fiftieth anniversary of Apollo 11 and the first human landing on the Moon, as well as the Apollo 9 and 10 missions preceding that landing, and the Apollo 12 landing that followed, also in 1969. But what will the Moon look like on the seventy-fifth anniversary of these missions? Some of that future has already begun to pan out in this new age of lunar exploration.

ACKNOWLEDGMENTS

SINCE STERLING PUBLISHING COMMISsioned this book for the fiftieth anniversary of NASA's first human Moon landings celebrated in 2019, acknowledgements must begin with the late US President John F. Kennedy for committing NASA to the ambitious goal on May 25, 1961. Along with recognition of the person at the top, I wish to acknowledge the nearly 400,000 people who worked on Project Apollo within the US government, industry, and academia. Although the names of most of these people are not remembered outside their own families and the places where they worked, the team included a particularly well-known group, namely the Apollo astronauts. Along with those lucky enough to become humanity's first representatives on another celestial body, particular acknowledgement is due to the three astronauts who died training for the mission that became known as Apollo 1: Gus Grissom, Ed White, and Roger Chaffee. In the course of researching and writing this book, I came to appreciate something that astronaut Frank Borman has pointed out in lectures ever since he led the accident investigation team following the Apollo 1 fire of January 1967, namely that the success of all the human Moon missions, occurring from 1968 to 1972, was possible because of the numerous changes that were implemented in the program in response to the Apollo 1 accident.

I am grateful to Sterling Publishing for giving me the opportunity to author this book, and especially for the work that the Sterling team put into the project beginning in the summer of 2017. In particular I thank the two editors who worked directly with me on the manuscript, Meredith Hale and Elysia Liang, as well as Michael Cea, Linda Liang, Christine Huen, Kevin Ullrich, Elizabeth Lindy, and Ellen Day Hudson for their contributions to the project. I am grateful to my literary agent, Regina Brooks of the Serendipity Literary Agency, for all the negotiations, advice, and other work that led to the book and helped it move along, but also for the preceding two years of talking up an idea for a totally different book, which led serendipitously to this book!

Acknowledgments also to historian Richard Carrier, PhD, for providing me some of his insight on science of the Hellenistic and Roman period, and to Michael D. Delp, PhD, of Florida State University for discussing by phone his research and insights regarding space radiation on the cardiovascular system.

FORMATION OF THE MOON

 HOW DID EARTH'S MOON COME INTO EXIStence, and when did it happen? Lunar samples and lunar probes have given us a precise answer to the "when" question. Analysis of rock fragments brought from the Moon's *Fra Mauro* highlands by astronauts on Apollo 14 suggests that the Moon is about 4.51 billion years old. The answer to the "how" question is more complex and always changing.

One idea, proposed by Édouard Albert Roche in 1873, is that the Earth and Moon formed next to each other in space. The two worlds contain vastly different proportions of iron, however; Earth contains much more iron, most of it concentrated in the planet's large, metallic core. Okay, let's try another one: What if the Moon formed far away and was captured later by Earth's gravity? Proposed by Thomas Jefferson Jackson See in 1909, the capture hypothesis also has problems, one being that ratios of isotopes of oxygen and certain other elements in the rocks of the two worlds match, almost like fingerprints, as if the Moon were made of the same materials as Earth's outer parts. Might Earth thus have flung a chunk of itself into orbit? George Darwin, son of naturalist Charles Darwin, proposed this scenario in 1879, but it too presents problems.

There is also the Giant Impact hypothesis. Proposed by William Hartmann and Donald Davis in 1975, the scenario is that a Mars-sized planet, which scientists call *Theia*—named for the mother of Selene, goddess of the Moon—struck Earth soon after both planets had differentiated into crust and core. Such a history fit well with findings from Project Apollo that the Moon lacked water and other volatiles (substances that boil easily). It also fit with findings from Apollo and the Lunar Prospector probe that the Moon has at most a very small iron-sulfur core, but the matching of the isotopes between the worlds and issues involving the Earth's spin and the Moon's orbit have led scientists to propose variations as well as entirely different scenarios. This includes a multiple small-impact hypothesis that was described in 2015 by Oded Aharonson and colleagues of the Weizmann Institute and that took shape in 2017.

SEE ALSO: Scientists Consider Lunar Origins (1873–1909), Return to the Moon (1971), Elucidating Lunar History (1970s–80s), Preparing for New Missions (2018)

Since the 1870s, scientists have proposed several hypotheses to explain the Moon's origin.

OCEAN TIDES

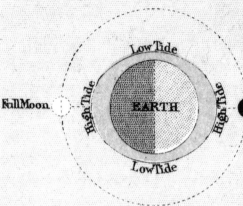

SPRING TIDES

Low Tide
High Tide
EARTH
High Tide
New Moon
Full Moon
Low Tide
SUN

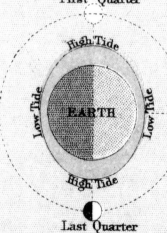

First Quarter

High Tide
Low Tide
EARTH
Low Tide
High Tide

Last Quarter

NEAP TIDES

SUN

MOON-EARTH PULLING BEGINS

WHEN THE MOON FORMED, A DAY ON Earth may have taken just four hours, and the Moon was at a fraction of its current distance. It would have looked huge in Earth's skies. But Earth's spin has slowed, while the Moon has moved farther away.

This cosmic dance comes down to *tidal forces*. Gravity between two objects decreases with distance, making the Moon's gravity stronger on the closer side of Earth and weaker on the opposite side. Earth is thus stretched, slightly in its rocky structure, and substantially in the more flexible oceans. This makes a bulge on the side of the planet facing the Moon, and another bulge on the side facing away from the Moon. Halfway between the two ocean bulges, perpendicular to the Earth–Moon axis, the oceans flatten. As Earth spins, this geometry causes two high tides, alternating with two low tides, each day. The difference between high and low tide changes because the Moon's distance changes over the course of each orbit, and because the Sun's gravity also produces tidal bulges. Solar tides strengthen or weaken the lunar tides, depending on the angle between the Moon, Earth, and Sun. But the lunar tides dominate, because the Sun is much farther away.

Because the water shift lags behind the gravity changes, Earth's rotation puts the Moon-facing water bulge slightly ahead of the Moon's position. Since water has gravity, the Moon is pulled forward, and the increased speed drives it farther from Earth, roughly four centimeters per year. As the ocean pulls the Moon forward, the Moon pulls back, slowing the Earth's spin.

Earth also causes tidal stretching of the Moon's rocky structure. Billions of years ago, this forced the Moon's spin to be the same duration as the Moon's orbit around Earth. This is called *tidal locking*. It's the reason why just one side of the Moon is visible from Earth. Billions of years from now, Earth *could* become tidally locked to the Moon, leaving the Moon visible from only one side of the planet, but the Sun will actually prevent this by swelling to engulf both worlds.

SEE ALSO: The Moon and the Origin of Earth Life (4.3–3.7 Billion Years Ago), Applying Math to the Lunar Orbit (2nd Century BCE), Improving Instruments Advance Lunar Astronomy (18th Century), Scientists Consider Lunar Origins (1873–1909)

This engraving from 1891 shows how the ocean bulges during a spring tide and neap tide. Spring tides occur when the Moon, Earth, and Sun are aligned, while neap tides happen when the Moon and Sun are perpendicular to each other. Both types of tides occur twice per lunar month.

THE MOON AND THE ORIGIN OF EARTH LIFE

SCIENTISTS ESTIMATE THAT BILLIONS OF years ago, the Moon could have orbited as closely as 25,000 kilometers (15,500 miles) on average from Earth, about fifteen times closer than its current distance. This made the Moon's tidal forces about 225 times stronger than they are today. Interaction between water and land was intense. Today, the coastline shifts by distances measured in meters or feet between high and low tide. Billions of years ago, the shifting coast would have been measured in kilometers or miles.

In recent years, scientists have identified microscopic structures, assemblages of minerals, and chemistry suggestive of microscopic life in Earth rocks dated as far back as 3.95 billion years. In one study, published in 2017, possible microfossils were described in rocks from Quebec, Canada, whose date range may extend to 4.28 billion years ago. Controversy surrounds the latter date, but microorganisms were established on Earth by 3.7 billion years ago. Analysis of the lunar surface shows that this time frame corresponds roughly to the end of a period of intense bombardment of the Moon, Earth, and other inner planets by rocks from space. This raises a question of whether life took hold on Earth because the crust cooled enough for life to thrive as bombardment subsided, or because the bombarding rocks carried biologically important molecules to the Earth's surface.

We do not know exactly how life emerged from non-living chemical systems. Scientists are not even certain that the origin of life took place on Earth at all. Microorganisms could have transferred through space to the early Earth within rock material that is frequently ejected into space by impact events on planets, moons, dwarf planets, and other bodies. Nevertheless, there are several plausible hypotheses about how life could have started from the chemistry present on our home planet.

In each chemical scenario, the presence of a system for concentrating various organic molecules turns out to be immensely helpful. Enormous tides caused by Earth's huge—and at that time very close—Moon would have met the requirements. Thus, it's plausible that the Moon stimulated the origin of life on Earth, although the jury is still out.

SEE ALSO: Moon–Earth Pulling Begins (4.5 Billion Years Ago), Applying Math to the Lunar Orbit (2nd Century BCE)

The close proximity of the Moon to the Earth billions of years ago generated huge tides. These tides may have concentrated organic chemicals that eventually allowed life to emerge on Earth.

IMPACTS CARVE INTO LUNAR CRUST

OBSERVING A FULL MOON FROM EARTH, you'll always see the same, distinct patterns of darker and lighter areas. People know these patterns as the "Man in the Moon," and by the designations *terrae* (Latin for "lands") for the lighter regions, and *maria* ("seas") for the darker regions; the largest sea is even called an "ocean," though none are actually bodies of water. The patterns don't change from our vantage point, because tidal locking keeps the same lunar hemisphere, the "nearside," pointing toward Earth continuously. Over each month, we actually see a little more than a hemisphere, about 59 percent of the lunar surface, because of the shape and inclination of the Moon's orbit. Occasionally, some wobble makes still more of the Moon's "farside" visible, but the bulk of the farside is visible only to individuals and probes that travel to the Moon and fly around it.

For humanity, the Moon's nearside patterns have been a constant, familiar presence, but only in the past few decades have scientists gotten a grip on what could have produced them. The short answer is big impacts from space rocks—meteoroids, asteroids, and comets. The exact dating is not yet certain, but sometime between 4.3 and 3.9 billion years ago two enormous impacts are thought to have carved out extremely large chunks from the crust. One of these carvings, called the South Pole–Aitken Basin (SPA), covers an area about 2,500 kilometers wide (1,550 miles) on the Moon's farside. The other carving, located on the nearside, is even bigger, with a width of roughly 3,200 kilometers (2,000 miles). Known as the nearside megabasin (NSB), this ancient carving of the lunar crust is centered on Oceanus Procellarum (Ocean of Storms) on the western part of the nearside. It also encompasses two seas called Mare Imbrium (Sea of Showers) and Mare Serenitatis (Sea of Serenity), plus parts of other maria. But this enormous section of the Moon did not acquire the notable shapes visible from Earth until it was altered further by more impacts and volcanism.

SEE ALSO: A Lunar Facelift (3.9–3.1 Billion Years Ago), Peak of Lunar Volcanic Activity (3.8–3.5 Billion Years Ago)

An artist's conception of a space rock impacting the surface of the Moon.

A LUNAR FACELIFT

SCIENTISTS DIVIDE LUNAR HISTORY INTO periods that relate partly to timing of impacts that have altered the Moon's surface by creating craters, larger holes called basins, and other features. A lunar feature is called Pre-Nectarian if it was formed prior to the carving of a basin called Nectaris. The latter was formed by a strong impact that is thought to have occurred roughly 3.9 billion years ago. One suspected Pre-Nectarian structure is the Tranquillitatis basin, corresponding to the Sea of Tranquility, where the astronauts of Apollo 11 would make history. Another is the Hipparchus crater, whose namesake, Hipparchus of Nicaea (c. 190–120 BCE), is considered one of antiquity's most important astronomer-mathematicians. Still another crater from this time period is named for Archimedes of Syracuse (c. 287–212 BCE), a mathematician-physicist who is remembered as the greatest mechanical genius of antiquity.

Formation of the Nectaris Basin marks the beginning of the Nectarian Period, which ended roughly 3.85 to 3.77 billion years ago, when another big impact carved the Imbrian basin, corresponding to Mare Imbrium, right smack in the middle of the nearside megabasin (NSB). Formation of the Imbrian basin also marks the beginning of the Moon's Imbrian Period. Lasting until 3.2 billion years ago, this period also was the setting for other impact events that formed other notable features, including the Serenitatis basin.

In lunar science, the term basin connotes a hole greater than 300 kilometers (roughly 185 miles) in diameter, as opposed to a crater, which is smaller than 300 kilometers. A megabasin is an enormous basin that has other basins carved within it. This is the case with the NSB, which contains the Imbrium and Serenitatis seas. This is because the basins of those seas were carved by impacts into the floor of the NSB. But the carving of these basins was only the first step in the process that would lead astronomers of the past to call them *seas*.

SEE ALSO: Impacts Carve into Lunar Crust (4.3–3.9 Billion Years Ago), Peak of Lunar Volcanic Activity (3.8–3.5 Billion Years Ago)

An image of the Moon taken from the Galileo in December 1992 shows some of its most prominent lunar basins.

PEAK OF LUNAR
VOLCANIC ACTIVITY

WHY DO THE MARIA ON THE MOON look darker than the highlands? After the Moon formed, its outer layer cooled and solidified as the lunar crust. Deeper down, molten rock, or magma, formed the mantle, which remained molten for billions of years. Sometimes, impacts that formed basins, or even large craters, were strong enough to crack through the crust, down to the molten mantle, creating volcanoes, through which magma could escape.

In its youth, the Moon was volcanically active, with two peaks of intense volcanism about 3.8 and 3.5 billion years ago. In 2017, scientists from NASA's Marshall Space Flight Center and the Lunar Planetary Institute published research conducted on volcanic glass samples brought from the Moon by astronauts in the 1970s. The study revealed that magma of the early lunar period was so loaded with carbon monoxide, gaseous sulfur compounds, and other volatile agents that for roughly 70 million years the Moon actually had an atmosphere—a true atmosphere, not the scant gathering of non-interacting particles that it has today.

When volcanic activity sent magma up through cracks produced by impact events, the magma flowed over the cracked basin as basaltic lava, similar to the lava that flows in Hawaii. Soon, the lava cooled and hardened into basalt, a type of igneous rock. High concentrations of iron made this particular type of basalt less reflective of sunlight compared with the rocks and dust of lunar highlands. This is why the lunar maria look dark. Although lunar volcanic activity began subsiding about 3 billion years ago, it was a gradual slowing that took 2 billion years, and some maria basalt surfaces are esti-mated to have formed only 1.2 billion years ago.

Maria cover about a third of the lunar nearside, but less than 2 percent of the farside—yet the farside has received just as much pounding from space. This asymmetry is due to the crust being much thicker on the farside, so it rarely has cracked deeply enough to release magma. Discovering why the crust is thicker on the farside than the nearside is one of many scientific motivations for returning humans to the Moon.

SEE ALSO: A Lunar Facelift (3.9–3.1 Billion Years Ago), First Pictures of the Moon's Farside (1959)

This sequence of images shows the progression of lunar volcanic activity in half-billion-year time increments. The red areas denote where lava has most "recently" erupted.

ERATOSTHENIAN PERIOD OF LUNAR GEOLOGICAL TIME

PRIOR TO THE SPACE AGE, ASTRONOMERS could distinguish older from newer regions of the Moon by counting craters. More craters means that a region solidified from molten rock farther back in time. But analysis of lunar rocks brought by astronauts has provided actual ages of certain regions. This has allowed calibration of crater numbers to estimate ages of lunar sites from which we do not have samples, and also crater-covered regions of other planets.

Lunar craters formed by impacts occurring between 3.2 and 1.1 billion years ago represent an intermediate phase of lunar history. Whereas craters and basins older than 3.2 billion years have accumulated smaller craters made by impacts within them, intermediate-aged craters tend not to contain other craters. Their time of formation is called the Eratosthenian Period, named for the quintessential intermediate-aged crater Eratosthenes.

Crater Eratosthenes is named for Eratosthenes of Cyrene (c. 276–194 BCE), chief librarian at the Museum of Alexandria, in Egypt, who earned the nickname "Beta," after the second letter of the Greek alphabet. This was because he was considered to be the second-smartest person in the world, second to Archimedes, but the astronomer Carl Sagan (1934–96) once remarked, "It seems clear that in almost everything Eratosthenes was 'Alpha.'" Sagan had multiple reasons for saying this, one being that Eratosthenes was a polymath credited with discoveries in multiple fields. In mathematics, he invented a kind of sieve, an algorithm for picking out prime numbers. He essentially invented geography. Lines of latitude and longitude come from Eratosthenes, who also is remembered for calculating the tilt of Earth's axis of rotation.

Most important to our story of the Moon is that Eratosthenes found a way to measure the size of the Earth. Since another scientist of the era, Aristarchus of Samos (c. 310–230 BCE), had computed the size of the Moon as a proportion to the sizes of the Earth and Sun and to the distances between all of these bodies, Eratosthenes's measurement allowed these proportions to be converted into absolute distances.

SEE ALSO: Library of Alexandria (Early 3rd Century BCE), Aristarchus Measures Lunar Diameter and Distance (3rd Century BCE), Quarter Phase Moon and Heliocentrism (3rd Century BCE), Eratosthenes Calculates Earth's Circumference (3rd Century BCE)

Eratosthenes Teaching in Alexandria (1635) by Italian painter Bernardo Strozzi (c. 1581–1644) depicts the librarian instructing an attentive pupil.

COPERNICAN PERIOD OF LUNAR GEOLOGICAL TIME BEGINS

WHEN AN IMPACT CARVES A NEW crater, cracks radiate outward from the crater rim, like the cracks on your car windshield after a pebble strike. These cracks, called rays, are eventually eroded by micrometeoroids—tiny particles from space—but this erosion takes billions of years. Persistence of bright rays identifies a crater as having formed during the Copernican Period. Running from 1.1 billion years ago to the present. The period is named for its quintessential crater, Copernicus.

Crater Copernicus is named, in turn, for Nicolaus Copernicus (1473–1543 BCE), one of the last famous astronomers to work without a telescope. Other Copernican period craters are named for other thinkers from history. One is Thales of Miletus (c. 624–546 BCE), remembered as the earliest individual in Western civilization to ponder how nature might operate through forces that were predictable and not subject to the whims of the gods. Another Copernican crater is named for Anaxagoras of Clazomenae (c. 510–428 BCE). For declaring that the Moon and Sun were not gods but objects, Anaxagoras stood trial in Athens and would have been executed had not the statesman Pericles come to his rescue. Thales and Anaxagoras were part of the Ionian Enlightenment that took place on the islands and western coast of modern-day Turkey, where the frontier of the Persian Empire came up against the fiercely independent Greek merchants.

That ancient intellectual awakening opened the door for later Ionians, including one astronomer and mathematician, the namesake of a particularly bright Copernican crater called Aristarchus. In the third century BCE, Aristarchus posited that Earth orbited the Sun, not the other way around, and he came to this conclusion through measurements, geometry, and calculations.

Jumping ahead two millennia takes us to two other astronomers, each with a Copernican crater designated in his name. One is Tycho Brahe (1546–1601), who lived after Copernicus, also a bit too early to make use of telescopes. The other is Johannes Kepler (1571–1630), most famous for his laws of planetary motion.

SEE ALSO: Thales Stops a War (6th Century BCE), Anaxagoras Stands Trial (5th Century BCE), The Moon Orbits Alone (1543), Moon and Sun Orbit Earth (1570s), A Dream of a Lunar Voyage (1581)

A photograph of Copernicus (the crater in the top center) with cracks called rays emanating from its rim.

IMPACT FORMS THE ARISTARCHUS CRATER

I N THE NORTHWEST REGION OF THE MOON'S nearside, a plateau rises up from the surrounding lava deposits that comprise the *Oceanus Procellarum*. The basalt rock that once formed from the lava is dark, but the plateau is bright. Named for Aristarchus of Samos, the plateau contains a large crater, also called Aristarchus. You can see it with the naked eye, because it is larger than the Grand Canyon, and its *albedo*—its reflectivity—is even higher, compared with the plateau. Due to its extreme brightness, Crater Aristarchus has been a favorite topic of conspiracy believers who imagine it serving a range of functions, from nuclear-fusion power generation to bases and portals for UFOs.

In reality, the crater is bright because of the mineralogy of the region, which was a focus of two Apollo missions in the 1970s and several robot probes in the 1990s and early 2000s. It will also be a target of future lunar missions, since we do not yet understand it. Maybe that's appropriate given the role of Aristarchus the astronomer, in humanity's earliest attempts to understand the universe and our place within it.

Aristarchus was not popular in the ancient world, because he did not ponder matters of existence, ethics, and the basis of knowledge that concerned the great philosophers of his time. Instead, he took careful measurements of the angles between the Moon, Sun, and Earth, timed the passage of the Moon through Earth's shadow during a lunar eclipse, and made calculations based on his data. His calculations turned him into the first person in history to explain that the Earth orbited the Sun, the first person to disprove an assertion by Aristotle of Stagira (c. 384–322 BCE; usually known simply as Aristotle) that Earth lay at the center of the universe—and he did not stop there. Because the stars did not seem to shift in the skies, as they should have shifted if Earth did indeed orbit the Sun, Aristarchus also proposed that the stars were so far away that they must be suns in their own right, perhaps even with worlds of their own.

SEE ALSO: Aristarchus Measures Lunar Diameter and Distance (3rd Century BCE), Quarter Phase Moon and Heliocentrism (3rd Century BCE), Applying Math to the Lunar Orbit (2nd Century BCE), Improving Instruments Advance Lunar Astronomy (18th Century)

Crater Aristarchus, located in the center left of this photograph, is large and bright enough to be seen from Earth with the naked eye.

LUNAR ASSISTANCE
FOR INTELLIGENT LAND LIFE

LUNAR TIDES MAY OR MAY NOT HAVE BEEN key to the emergence of life, but the Moon helped shape life once Earth's first living things came into existence. Like the poles that tightrope walkers hold to stay balanced, the Moon stabilizes Earth. Our planet's rotation axis is tilted 23.44° with respect to the plane of its orbit around the Sun. That tilt accounts for seasons over the course of each year, but the angle has not always been 23.44°. Earth wobbles a bit as it spins, like a dreidel rotating too slowly. Earth's wobbling shifts the rotation axis between roughly 21.5° and 24.5° over thousands of years. The wobbling contributes to the coming and going of ice ages and the shifting of the positions of stars as seen from Earth.

For most of Earth's history, life consisted of only singled-celled organisms, which still account for most Earth life. Single-celled organisms have a tendency to be more resilient and tenacious compared with multi-celled organisms, particularly large, complex creatures such as plants and animals, which colonized land roughly 440 million years ago. Several times in geological history, complex life, big life, came close to extinction. Notoriously, in the Permian Extinction that happened 252 million years ago, most of the land life visible to the naked eye, and a large chunk of life in the sea, was killed off.

Much later, after brains had evolved beyond chimpanzee-size, as humanity was becoming recognizable a few million to 1.5 million years ago, genetic studies suggest that the population of humanity's ancestors approached extinction more than once. Similar near-extinctions also threatened humans more recently, in the ice ages of the Pleistocene epoch, in all cases the climate playing a huge role. In all this time, however, the Moon kept the planet's tilt within that narrow range of 21.5° to 24.5°. Without such stabilization, the wobble would have been far greater, the changes of climate far more extreme. Intelligent life might not have had the time it needed to come into existence, especially on land. Humans might not have lasted long enough to create civilizations and, ultimately, the science and technology that would teach us the Moon's value.

SEE ALSO: The Moon and the Origin of Earth Life (4.3–3.7 Billion Years Ago), Applying Math to the Lunar Orbit (2nd Century BCE), Improving Instruments Advance Lunar Astronomy (18th Century), Return to the Moon (1971)

In this diagram, the red line running through the Earth's poles shows the Earth's axis of rotation. Over thousands of years, Earth has wobbled, shifting the rotation axis between roughly 21.5° and 24.5°. Its wobbling would be more extreme if not for the Moon.

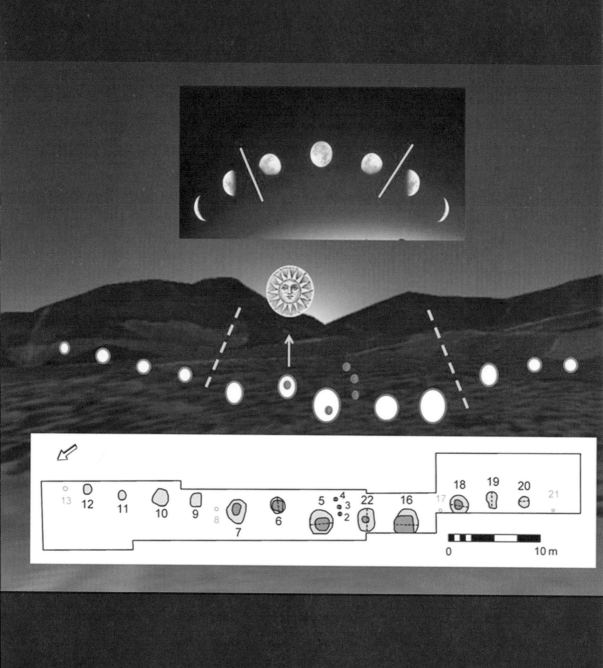

13 12 11 10 9 8 7 6 5 4 3 2 22 16 17 18 19 20 21

0   10 m

MESOLITHIC LUNAR CALENDAR

WATCHING THE SKIES, STONE AGE observers would see the same lunar phases that we know today, in cycles lasting about 29.5 days. In each cycle, the Moon would widen gradually from a narrow crescent to a full circle, and then narrow again until it was no longer visible. Those who kept watching lunar cycles also would notice that the progression of seasons—winter to spring to summer to fall and back to winter, what we call a year—takes somewhere between 12 and 13 lunar cycles. It happened year after year, and so people started using the Moon as the basis of a calendar.

The earliest known example of a lunar calendar was constructed in Aberdeenshire field, Scotland, around 10,000 years ago. The era is what archaeologists call the Mesolithic, based on the emergence of new technologies that helped people to transition from the hunting and gathering that had characterized the Paleolithic era to the agriculture and more permanent settlements that would characterize the Neolithic Revolution that would follow.

Although people must have been using various lunar calendars throughout the world, the Aberdeenshire calendar has survived to our time because it was a massive physical structure that was literally carved into the Earth. Along an arc spanning about 165 feet (50 meters), Mesolithic people carved twelve pits, and shaped the pits to resemble the various lunar phases in the order that the phases occurred in the sky. To be useful over time, this arc of pits had to be synched with the solar year. The builders achieved this by aligning the center of the arc with a notch between hills in the landscape that lined up perfectly with the rising Sun on one day of the year, the shortest day, the winter solstice. Seeing the Sun through that notch each year, the people of Aberdeenshire would know to mark the next new moon as the start of the first month of a new year.

SEE ALSO: Complex Lunar Calendar Systems (18th–17th Centuries BCE), Nabonassar Standardizes the Lunar Calendar (747–734 BCE)

This diagram shows how the twelve pits found in Aberdeenshire, Scotland, correspond to the phases of the Moon. The pits vary in depth and are arranged in an arc. A notch between the hills in the landscape lined up with the rising Sun on the day of the winter solstice, so it could be used to reset the lunar calendar when it became out of sync with the seasons.

HUMANITY'S FIRST AUTHOR

B Y 2300 BCE, THE SUMERIANS OF SOUTH-ern Mesopotamia had a well-developed cuneiform writing system that was sophisticated enough to convey complex ideas and emotions. We know this because of writings preserved on clay tablets from Enheduanna, a lunar priestess and poet who flourished from 2285 to 2250 BCE.

Encompassing the land between the Euphrates and Tigris rivers, Mesopotamia was the site of Earth's first empire. It started with the conquest of Sumerian city-states by Enheduanna's father, Sargon of Akkad (who reigned from c. 2334 to c. 2279 BCE). According to legend, Sargon was Sumerian by birth but illegitimate. For protection, his priestess-mother had put the infant into a reed basket afloat in the river, where Akkadians, a more northern people, rescued and educated him. Sargon's birth story may have provided him some legitimacy in Sumerian cities, but it was not enough, so he also made use of religion.

Mesopotamian mythology portrayed the moon-god—sometimes called Nanna and sometimes Sin—as leader of a new generation of gods that had inherited their dominion legitimately. Thus, in an effort to associate himself with the moon-god, Sargon appointed Enheduanna as high priestess of Nanna's temple in the Sumerian city of Ur. As lunar high priestess, Enheduanna could govern the south, thereby helping her father to centralize political power in his northern city of Akkad. But Enheduanna also spent her time as a scribess of poetry. Most notable are her *Sumerian Temple Hymns,* a collection of poems dedicated to the daughter of moon-god Nanna-Sin, Inanna, goddess of love, fertility, and beauty, but also of war, political power, and the planet Venus. In the course of writing, she conveyed her personal dreams, but she also did something that no previous scribe had ever done: she identified herself by name. In doing so, Enheduanna connected Earth's Moon to humanity's ongoing drive to express itself.

SEE ALSO: Lunar Cults in the Bible (900–700 BCE), Earliest Mention of Selene (c. 7th Century BCE)

The Disk of Enheduanna depicts Enheduanna (second figure from the left), the high priestess of the Mesopotamian moon-god, presiding over a ceremony with three priests.

龍

MOON MEETS THE SUN OVER CHINA

HUMANITY'S HISTORICAL RECORD HAS NO shortage of anecdotes about kings, generals, or even whole nations getting spooked by what they thought to be the disappearance of the Sun or Moon. One notable example took place under the reign of Chung K'ang (2159–2146 BCE) in China. Apparently, the king was annoyed one day during a solar eclipse—what later would be understood as the Moon moving between the Earth and Sun, blocking the Sun's light—but not because the skies had gone dark. What actually irked him was the noise of drumbeats coming from a crowd of people who were trying to end the eclipse by scaring away "the dragon that was eating the Sun."

Chung K'ang may have forgiven his people when the Sun returned, but he had no sympathy for his court astronomers, who had failed to predict such a huge astronomical event. They were decapitated on the spot, perhaps partly to inspire any young, budding astronomers to do a better job of predicting celestial events. The plan didn't work so well, if an ancient document called the *Shu Ching* gives an accurate account of what happened when another solar eclipse occurred several years later, in 2134 BCE. By this time, a new ruler sat on the throne, but he continued the policy of capital punishment for court astronomers who failed to warn their ruler of a coming eclipse.

Court astronomy was a dangerous profession. Astronomers didn't have enough information to calculate future eclipses based on records from the past, but apparently they had an inkling of the underlying mechanics. We know this because the *Shu Ching* record of this second eclipse doesn't implicate a dragon for devouring the Sun: instead, it tells us, "the Sun and Moon did not meet harmoniously." From this we can deduce that by the year 2134 BCE, astronomers in ancient China did know that a solar eclipse had something to do with the Moon and Sun meeting up. They were on their way toward realizing that solar eclipses were caused by the Moon crossing in front of the Sun.

SEE ALSO: Assyrian Eclipse (763 BCE), Thales Stops a War (6th Century BCE)

Many in ancient China believed that solar eclipses were caused by a dragon eating the sun.

SUMERIAN LUNAR CALENDARS

To earn a living through farming, inhabitants of ancient river civilizations had to predict when the land would flood and dry, so they needed a way to reckon time with great precision. In Egypt, the priests paid particular attention to Sirius, a bright star whose rising just ahead of the Sun began at the same time of year that the Nile River began flooding the land. For the Sumerians in Mesopotamia, flooding of the Euphrates and Tigris rivers matched the schedule of no particular star, so their calendars relied entirely on the Moon and Sun.

The *Enuma Anu Enlil* is a collection of about 70 tablets discovered in the ruins of the Assyrian city of Nineveh, in northern Mesopotamia. Although dated to the seventh century BCE, two of the tablets mention a couple of lunar eclipses occurring some fourteen centuries earlier, around 2200 BCE. This was a time when a line of Sumerian kings—the Third Dynasty of the city of Ur—were asserting themselves, as the descendants of Sargon of Akkad were losing hold of the south. Sumerian astronomers were thus making astronomical recordings in this early period, possibly as part of an effort to synthesize calendar systems accurate enough to support a society that was growing increasingly sophisticated. This, after all, was a time of expansion in industry and trade and implementation of an early law system, called the Code of Ur-Nammu, so timekeeping needed to be accurate.

But creating a solar year consisting of a whole number of days, or a lunar year consisting of either 12 or 13 moons, could suffice for just a few years. After a generation or so, the winter would be coming when the calendar says it should be summer and vice versa. To mark time accurately enough to support urban life under Ur's Third Dynasty, the timing of both solar and lunar movements had to be understood with immense precision. And so, by the mid-21st-century BCE, formalized lunar calendar systems were spreading, not just in Ur but throughout Sumeria.

SEE ALSO: Complex Lunar Calendar Systems (18th–17th Centuries BCE), Nabonassar Standardizes the Lunar Calendar (747–734 BCE)

One of the seventy clay tablets that comprise the *Enuma Anu Enlil*. The tablets were written in cuneiform, wedge-shaped characters used for writing in Mesopotamia.

COMPLEX LUNAR CALENDAR SYSTEMS

I N THE EIGHTEENTH CENTURY BCE, THE Sumerian cities came under the sway of King Hammurabi of Babylon (reigning c. 1792–1750 BCE). Hammurabi is remembered best for his Code of Laws, which he imposed on those living within his expanding reign; but the flourishing international trade of this first Babylonian Empire also stimulated increasing civic organization at the local level. It was during this time period that the Sumerian city of Nippur established the lunar calendar for official use. The year on this calendar began in the spring, on the first day of the month of Nisanu.

Watching the Moon from Earth, our ancestors counted 29.5 days on average for each new moon to wax through crescent and gibbous shapes to a full moon and then wane until it was no longer visible. Known as the synodic month, the lunar cycle as seen from Earth is longer than a sidereal month, the time for the Moon to orbit 360°, which takes 27.3 days. The extra 2.2 days are there because Earth's motion around the Sun requires the Moon to orbit more than 360° to show us all phases.

But using half days on a calendar was not practical, so the ancient Sumerians alternated between 29-day and 30-day lunar months. This left the Moon invisible from Earth at the start and end of each month and full at the midpoint of each month. This system was useful initially for dating the regnal years of Hammurabi and his successors, and of local governors. But since a 12-month year came out to just 354 days, over the course of decades the lunar calendar fell out of sync with the 365-day solar calendar that was used for agriculture. To manage the problem, different cities began adding a thirteenth lunar month from time to time. This solved part of the problem; but having so many varying calendars made for confusion in an empire that was struggling to hang on to vast areas of the known world.

SEE ALSO: Sumerian Lunar Calendars (22nd–21st Centuries BCE), Nabonassar Standardizes the Lunar Calendar (747–34 BCE)

A tablet portraying Hammurabi showing adoration to Shamash, the Mesopotamian sun god.

LUNAR CULTS IN THE BIBLE

BIBLICAL WRITINGS SUPPORT THE proposition that ancient Israel was a setting for conflicts between cults devoted to different celestial objects, including the Moon. One strand of biblical text—known in the field of biblical source criticism as the *Yahwist* text—stands out for highlighting heroic female protagonists who follow a kind of nature-fertility religion whose goddesses have known associations with either the Moon or the planet Venus.

Writing in support of the kings of Judah during the ninth or eighth century BCE, when there were two Israelite monarchies (Samaria/Israel in the north and Judah in the south), the Yahwist writer names the "Book of Yashar" as the source of a story in which the character Joshua stops the Moon from moving from the Valley of Ayalon, a region that may have held a lunar temple. Meanwhile, the Sun does not leave Gibeon, a town where later biblical authors describe King Solomon as offering sacrifices to a version of Yahweh who has solar qualities.

The later authors condemn Solomon for worshiping various celestial objects, but still another biblical source text recounts a battle at Gibeon between the houses of David and Saul with conspicuous numbers of deaths on each side. Saul's house loses 360 men—the number of days in the Egyptian solar calendar. David loses 19—the number of years in each cycle of a hybrid lunar–solar calendar that began with King Nabonassar of Babylon (governed 747–734 BCE). Was the story allegorical of struggles between cults, calendars, and civilizations?

As for the female protagonists, a fascinating study of an obscure Yahwist text by Professor Ilana Pardes of Hebrew University hints that Moses's wife was a de-deified version of a Venus-associated goddess. Meanwhile, the Yahwist text describes the character Joseph dreaming of his mother Rachel as the Moon, but she may have been a lunar goddess originally, hailing from Haran, where the moon-god Sin had a sanctuary, and where the moon-goddess *Nikkal* was revered. Spelled in Canaanite script, the names Nikkal and Rachel are nearly identical, while, in Hebrew, the name of Rachel's father, *Laban*, literally means Moon.

SEE ALSO: Nabonassar Standardizes the Lunar Calendar (747–34 BCE), Earliest Mention of Selene (c. 7th Century BCE)

A photo of Rujum en-Nabi Shu'ayb, taken in 2015. Located in Israel, this ancient monument consists of stones arranged into the shape of a crescent moon and dates between 3050 and 2650 BCE. It is believed to serve as a shrine to a lunar deity and built by inhabitants of an Early Bronze Age community.

ASSYRIAN ECLIPSE

ASTRONOMY AND MATHEMATICS WERE invented together, each helping the other to grow more sophisticated over time, but initially astronomy was not a science set on understanding how the universe works. Rather, for thousands of years, astronomy was astrology, an attempt to predict actions of the gods on Earth, based on their motion through the skies.

Although their god-status varied among societies by the eighth century BCE, the Sun and the Moon were seen almost universally as indicators of whether things were going good or bad. The Sun going dark, the Moon going dark or red, the appearance of what today we call a comet—these were interpreted as omens. In the eight century, a plethora of superstitions involving celestial lights held as much sway over society in Mesopotamia as they did anywhere else. Thus, a full solar eclipse observed in the Assyrian city of Nineveh in the year 763 BCE was seen as a fateful omen throughout the land, even by people in the highest ranks of power. Kings who lost battles blamed it on the eclipse. Kings who won battles saw the return of the Sun's light as a sign that the gods had chosen them. Religious preachers who were being ignored suddenly won large followings, if their timing was favorable. It has been hypothesized that aspects of the biblical story of Jonah could have been related to the famous Nineveh eclipse.

But amidst all of the worrying and boasting surrounding the arrival of the Nineveh eclipse, Mesopotamian astronomers were at work recording the event, just as their ancestors had recorded a previous solar eclipse 300 years earlier and numerous other solar eclipses and lunar eclipses. In Mesopotamia, astronomers based southward of Assyria, in Babylonian lands, were most diligent about recording celestial observations.

SEE ALSO: Moon Meets the Sun Over China (22nd Century BCE), Thales Stops a War (6th Century BCE)

This illustration shows King Ashur-dan III (governed 772–775 BCE) watching the Nineveh eclipse, considered to be an omen that indicated the displeasure of the gods.

NABONASSAR STANDARDIZES THE LUNAR CALENDAR

I F ANY EIGHTH-CENTURY LEADER COULD BE called "pro-science," it was King Nabonassar, who governed Babylonia from 747 to 734 BCE. Nabonassar was Sumerian, so he knew the practice of adding extra months to the lunar year, but he wanted all cities to follow a standardized schedule. Combing through astronomical records and making calculations, Nabonassar's astronomers realized that 235 lunar cycles corresponded almost exactly to 19 solar years. This enabled a hybrid lunar–solar calendar that repeated in 19-year cycles with twelve of the years containing 12 months and seven years having a thirteenth month added at a specific time. The Greek astronomer Meton would implement the same system in Athens in 432 BCE. Still many centuries later, the Jewish astronomer Samuel of Nehardea (c. 177–257 BCE) would transform the Babylonian calendar into the Jewish calendar with the same 19-year cycle that had come into use in Nabonassar's time.

Meanwhile, Nabonassar's astronomers also learned that over the course of 223 synodic months—18 years, 11 days, and 8 hours—a solar eclipse would repeat somewhere on Earth. After three cycles—54 years and 34 days—it would repeat in almost the same geographic location. Known as the Saros cycle, this pattern also incorporated the more frequent lunar eclipses, and it meant that future eclipses were predictable.

King Nabonassar's astronomers were also astrologers who were expected to use celestial movements to foretell the future. But the astronomers' discovery that eclipses were predictable implied that the movements of the Moon and Sun were simply physical phenomena with no mystical significance. Nabonassar's astronomers probably never discussed such implications, but Greek thinkers would soon take notice and set astronomy on a path that ultimately would leave astrology in the rear-view mirror. Meanwhile, Babylonia was centuries ahead of Greece in mathematics, and had a knack for taking precision measurements of nature that would become as vital to scientific progress as the Greek talent for creating models to explain nature's workings.

SEE ALSO: Sumerian Lunar Calendars (22nd–21st Centuries BCE), Lunar Cults in the Bible (900–700 BCE), Seeing the First Sliver of a New Moon (11th Century)

This tablet fragment shows a remnant of a hybrid lunar-solar calendar, created based on observations from the astronomers working during King Nabonassar's reign.

EARLIEST MENTION OF SELENE

ARCHEOLOGICAL STUDY HAS REVEALED fragments of writing dating close to the time period of the Greek poet Sappho (who may have lived c. 630–570 BCE, but certainly lived subsequent to the mid-eighth century BCE, when the Greeks had emerged from their Dark Age) referring to the lunar goddess Selene. One of the most famous lunar goddesses of human culture, Selene had been a subject of oral tradition for many centuries. By the time of Sappho, Selene was considered not merely a lunar goddess, but one and the same with the Moon itself. A few centuries later, the Greeks would come to associate Selene with Artemis, twin sister of Apollo, and with a goddess of witchcraft called Hecate. This would transform Artemis and Hecate into lunar goddesses as well, but unlike Selene they were not personifications of the Moon.

As with so many of the Greek gods, poetry about Selene centered on romantic life. Thus, a well-known story concerns her love for Endymion, a human, with whom she had fallen in love. Being immortal, Selene did not wish to outlive her mate. This led her to consider asking Zeus to grant Endymion eternal life; but her sister Eos, goddess of the dawn, had asked Zeus to do the same for her human husband. Zeus had granted the request, and as a result Eos's husband had grown immensely old—so old that he had shrunk and shriveled into a grasshopper, for Eos had forgotten to ask Zeus also to grant him eternal youth. In order to avoid her sister's mistake, Selene requested eternal youth for Endymion; but to keep Endymion youthful, Zeus had to put him into an eternal sleep. Selene gave Endymion fifty daughters, but for all eternity, she was married to a sleeping man.

Scholars have expressed doubt as to whether the Greeks actually believed their myths, and the story of Selene and Endymion provides a good reason for such doubts. Selene's story persisted in any case, demonstrating that mythology surrounding the Moon was just as important to the ancient Greeks as to the various other cultures that shared their world.

SEE ALSO: Humanity's First Author (23rd Century BCE), Lunar Cults in the Bible (900–700 BCE)

In *Selene and Endymion* (c. 1770) by Italian painter Ubaldo Gandolfi (1728–81), the Greek moon goddess Selene watches her husband Endymion in the company of Eros.

MAP V.

GREECE
AND
HER COLONIES.

From Thalheimer's General History, by perm

Scale of Miles.

0   25   50   150

VICINITY
OF
ATHENS.

BEGINNINGS OF
NONRELIGIOUS ASTRONOMY

THE TWENTIETH-CENTURY PHYSICIST Richard Feynman (1918–88) taught that modern physics depended on a Babylonian approach to mathematics, wherein experience in expressing problems mathematically enables one to generalize to discover laws of nature. In contrast is the approach of Euclid, a Greek thinker who employed basic rules of logic to derive more complex theorems from fundamental truths that we call axioms. Euclid flourished around 300 BCE, but his derivation approach had its roots a few centuries earlier, when Greek thinkers first ventured to comprehend nature in a qualitative sense.

To make real progress toward a qualitative understanding of the Moon, Sun, and planets, Greek astronomers would have to embrace quantitative methods. They would need the Babylonian approach to astronomy, which included both practical mathematics and massive data collection from tedious observations of the sky. They would have to combine those data with a qualitative, conceptual approach that was distinctly Greek.

That conceptual approach had its origins among the Ionians of the sixth century BCE. Ionia consisted of islands and cities along the western coast of modern-day Turkey. It was the home of Greek peoples who began thinking that nature was knowable and predictable, for it operated independent of the whims of the gods. Initiated by Thales of Miletus, this demystified world view liberated the Ionians to propose physical models of natural phenomena. This was vital to the development of science.

But Thales was opposed by a different group of Greek philosophers, based on the other side of the Greek world, the colonies of southern Italy—Magna Grecia. Some Magna Grecian philosophers were ahead of the game in realizing that the Moon was spherical, not a disk. One group of Magna Grecians would introduce the mathematics needed to jump-start Greek astronomy. But the Magna Grecians were not scientific. They posited mysticism and disdained experimentation. The most influential mystical school also hid and suppressed inconvenient discoveries. Ironically, the founder of that influential mystical group, Pythagoras (c. 570–495 BCE), was born on Samos, an island in the heart of Ionia, and began his career as one of Thales's students.

SEE ALSO: Thales Stops a War (6th-Century BCE), Spherical Harmony (6th Century BCE)

A map of ancient Greece, created in the late nineteenth century.

THALES STOPS A WAR

HAD THERE BEEN A NOBEL PEACE PRIZE 2,600 years ago, it would have gone to Thales of Miletus. Born at a time when the rise of Babylon was facilitating an unprecedented era of intellectual activity, Thales developed an interest in mathematics and astronomy. Ionia was part of the Lydian kingdom, but Thales traveled to Babylon, or he got access to Babylonian astronomical writings. One way or another, he learned about the Saros cycle of lunar and solar eclipses, which dovetailed with his idea that the gods had no influence over nature.

Knowing that the commanders in an ongoing war between Lydia and one of its enemies, Media, were not so enlightened, Thales warned that the gods demanded a truce. To prove it, they would darken the Sun in 585 BCE, on a particular spring day when Thales's calculations, based on the Saros cycle, suggested that there should be a solar eclipse. In utilizing the Saros cycle, Thales actually was calculating the Moon's movement in front of the Sun, although he didn't know it. Nevertheless, when an eclipse actually happened, the war ended, and Thales earned a reputation in Ionia. Students flocked to study nature with him in Miletus. One such student was Pythagoras, who traveled by boat from the island of Samos.

Miletus at the time was a wealthy port city, whose Greek residents had no loyalty for empirical rulers. They were independent and open to new ideas, maybe because they heard so many different ideas from sea traders. And so Thales and the other Milesians started thinking about the world in very radical ways. They proposed that earthquakes resulted from large ocean waves striking land, for instance, and that the continents had formed from the seas' depositing silt. Their explanations would prove wrong, but what mattered more was how their thinking differed from all the rest of humanity. Their ideas were testable, capable of being disproven, because they did not involve gods who might be hiding. Thales and the Ionians who came after him were different because they decided that nature was knowable, through observation and analysis—and, for some of them, also through experimentation.

SEE ALSO: Moon Meets the Sun Over China (22nd Century BCE), Assyrian Eclipse (763 BCE), Nabonassar Standardizes the Lunar Calendar (747–34 BCE), Beginnings of Nonreligious Astronomy (6th Century BCE)

A drawing of Thales by Dutch painter and engraver Jacob de Gheyn (1616), depicting the Greek astronomer. The Dutch led the world in lens crafting in the seventeenth century. Thus, anachronistically, Thales is wearing a pair of glasses. Thales predicted a solar eclipse in 585 BCE.

SPHERICAL HARMONY

PYTHAGORAS OF SAMOS IS THE EARLIEST person known to say that the Moon was spherical. He may have been motivated initially by an observation, such as the curvature of the lunar terminator, the line separating the lit and unlit sides of the Moon. Pythagoras was a student of Thales, after all, and recognized, ahead of his contemporaries, that the Morning Star and Evening Star were the same object, the planet Venus. This realization depended on observation, although Pythagoras would come to reject observation, insisting that the universe could be understood through pure thought.

After living in Egypt for many years, and later Babylon, Pythagoras introduced the theorem that the area of a square formed from the hypotenuse of a right triangle equals the sum of the areas of squares formed from that triangle's two shorter legs. It was not his idea. Egyptians and Babylonians had been putting the concept to practical use for ages, in Babylon even expanding it into trigonometry. Nevertheless, by introducing the theorem to the Greeks, Pythagoras would enable mathematics that future generations would need in making sense of nature.

For Pythagoras, however, mathematics was not a tool, but a religion. His concept of the spherical Moon was part of a mystical outlook of a cult—the Pythagorean brotherhood—that he founded in the Croton colony in Italy. The Pythagoreans posited a "harmony of spheres" in the heavens, with the Moon and planets being not merely spherical, but perfectly spherical, and moving in perfectly circular orbits, each sphere generating a particular musical note. Along with a disdain for observation, Pythagoras routinely suppressed discoveries that conflicted with his perfection concept. This included the discovery, by a student, that the square root of two was irrational, not expressible as a ratio of two whole numbers.

There was a rumor that Pythagoras had murdered the student to silence him, but Pythagoras didn't need violence to promote his beliefs. Soon, the philosopher Plato (c. 427–347 BCE) would embrace Pythagorean mysticism, complete with perfect spheres, circular orbits, disdain for observation, and all the trimmings needed to thwart the emergence of science for centuries to come.

SEE ALSO: Beginnings of Nonreligious Astronomy (6th Century BCE), Thales Stops a War (6th Century BCE), Earth's Curved Shadow on the Moon (c. 350 BCE), Heavenly Perfection Corrupted (350 BCE)

An eighteenth-century etching of Pythagoras, after Italian painter Raphael's (1483–1520) rendition of the Greek thinker in *The School of Athens* (1511).

ΑΝΑΞΑΓΟΡΑΣ

ANAXAGORAS STANDS TRIAL

WHEN ASKED WHAT MADE LIFE WORTH living, Anaxagoras of Clazomenae replied "the investigation of the Sun, Moon, and stars." How fitting, as he also was the first person to explicitly say that the Moon was not a deity. Instead, it was a rock that Earth had flung into space, and that merely reflected light from the Sun, which also was not a deity, but a burning rock. These ideas enabled Anaxagoras to explain what we call eclipses. Lunar darkening must result from the Moon, Earth, and Sun lining up, such that the Moon fell in Earth's shadow, Anaxagoras proposed. Similarly, he reasoned that darkening of the Sun must result from the Moon passing directly in front of the Sun. He was correct, and yet, toward the end of his life, Anaxagoras is said to have stood trial in Athens on charges of impiety and sentenced to death.

How could this happen, in the era of the great statesman Pericles (c. 495–429 BCE), of all times, when Athenian democracy was thriving, when people embraced new ideas? Tolerance and cooperation certainly had been the trend over Anaxagoras's lifetime. During his youth, he had seen an alliance between Athens and Sparta drive the Persian Empire from Ionia.

Anaxagoras had then traveled to Athens, where he was both a teacher and a friend to Pericles. Attempting to resolve a complex debate over the essence of existence between followers of the mystic Parmenides of Elea (c. 515–450 BCE) and the naturalistic Ionians, Anaxagoras posited that all matter contained seeds of everything that could exist. This led him to propose that living seeds could spread through space, planting life on multiple worlds.

Pericles, however, made errors that later would lead to the Peloponnesian wars against Sparta. To wound him politically, Pericles's opponents went after his friends, and historians are fairly certain that there was impiety legislation on the books that could be used against Anaxagoras, just as it would be used later to prosecute Socrates. Fortunately, Pericles managed to free Anaxagoras from jail before the sentence could be carried out, but the latter spent the rest of his life in exile.

SEE ALSO: Copernican Period of Lunar Geological Time Begins (1.1 Billion Years Ago), Greeks Understand Lunar Phases (5th Century BCE), Eastern Astronomers Keep Looking Up (500–800), Scientists Consider Lunar Origins (1873–1909)

An illustration of Anaxagoras, who accurately described how solar eclipses result from the Moon passing directly in front of the Sun.

5th Century BCE

45

GREEKS UNDERSTAND LUNAR PHASES

ACH MONTH, THE MOON WIDENS FROM A narrow crescent through a half circle and gibbous phases to a full moon, and then shrinks back to darkness. To understand why the Moon goes through phases, one must recognize that moonlight is reflected sunlight, and also that the Moon is a sphere whose position changes continuously with respect to the Earth and Sun. Being spherical, the Moon is always lit on half of its surface, but we see the entire lit side only when the Sun, Earth, and Moon are nearly lined up with the Moon and Sun at each end of the line. If it's a perfect line, Earth's shadow falls on the Moon, so we see a lunar eclipse; but most months, this doesn't happen. On the other hand, when the Earth, Moon, and Sun are similarly nearly lined up but with the Moon between the Sun and Earth, the Moon is dark to us—a new moon—because the entire lit side faces away from us. And if the Moon passes exactly between the Earth and Sun, there is similarly a solar eclipse somewhere on Earth. Most months this does not happen, but each new moon looks dark to us because the Moon is in almost the same direction as the Sun.

As the angle between the Sun, Earth, and Moon increases, we see increasingly more of the lit lunar hemisphere; we see a waxing Moon. We see a waning Moon as the angle decreases in the second half of the lunar month. In the fifth century BCE, Anaxagoras of Clazomenae grasped lunar and solar eclipsing, because he knew that the Moon did not make its own light. It's not clear whether he expanded the concept to explain lunar phases, but two of his distant contemporaries, Empedocles (c. 490–430 BCE) and Parmenides (c. 515–450 BCE), also acknowledged that moonlight was reflected sunlight, Parmenides describing poetically the bright side of the Moon "always facing the light of the Sun." This implies a mechanistic understanding of lunar phasing floating around the Greek world of the 400s bce, a century before Aristotle of Stagira would jot it down in a more prosaic, explanatory manner.

SEE ALSO: Earth's Curved Shadow on the Moon (c. 350 BCE), Anaxagoras Stands Trial (5th Century BCE), Heavenly Perfection Corrupted (c. 350 BCE)

This seventeenth-century engraving shows the twenty-eight phases of the Moon that occur in one lunar month.

EARTH'S CURVED SHADOW ON THE MOON

T HE MOON PLAYED A ROLE IN HELPING ancient people realize that they were living on a curved Earth. Unequivocal statements about the Earth being a sphere date back at least to Pythagoras, who revered the spherical form religiously. It's not clear whether Pythagoras's thinking about the Earth was motivated—at least initially—by observations of nature, despite his eventual rejection of observation as a pathway to understanding the universe. But given the importance of sea travel to the economy of the island of Samos, where Pythagoras grew up, and to the rest of the Greek world, it seems unlikely that nobody of sixth-century Ionia would have deduced the curvature of the Earth's surface. Seeing any ship approaching a harbor from a distance, people would have noticed that the mast of the ship was visible first and grew taller before the hull was visible, as if the ship were rising as it came closer. Sailors on a ship, similarly, would see the top of a mountain before seeing the lower parts of an island as they approached, and the reverse would happen for ships moving away from land. This is common-sense evidence for the curvature of the Earth, evidence that you can see for yourself today if you watch boats from the beach.

There were other bits of evidence available to ancient peoples, too, but it was not until the fourth century BCE, as far as we know, that somebody organized the evidence together and presented it systematically to make a strong case that the Earth was indeed a sphere. That somebody was Aristotle of Stagira. Along with recounting the observations made concerning the masts of ships, Aristotle noted that the altitude of the star Polaris (the "North Star") and the constellations close to it would change as one traveled great distances between north and south. Things would not look this way from a flat Earth. As another line of evidence, he also pointed out that during a lunar eclipse, astronomers always saw the Earth's shadow on the Moon with a curved edge. This meant that Earth's surface was curved from all directions.

SEE ALSO: Heavenly Perfection Corrupted (c. 350 BCE), Quarter-Phase Moon and Heliocentrism (3rd Century BCE), Seeing the First Sliver of a New Moon (11th Century)

A photograph of a lunar eclipse taken in the early morning of January 31, 2018, showing the curved shadow that Earth casts on the Moon.

HEAVENLY PERFECTION CORRUPTED

COMPARED WITH HIS TEACHER PLATO OF Athens, Aristotle of Stagira was more scientific. Whereas both men set out to synthesize the philosophies of mystics such as Pythagoras and Parmenides with the naturalism of the Ionians, Plato leaned toward the mystics, while Aristotle was mostly friendly to the Ionians. This was especially true when it came to Ionian *empiricism*, the idea that knowledge must be obtained through experience by way of the senses. Observing the deposition of silt by the Nile River, the Ionian Thales of Miletus hypothesized that the world's entire land mass had formed from a primordial sea through a similar process. Observing differences between baby fish and humans, and looking at fossilized skeletons, Thales's student Anaximander of Miletus (610–545 BCE) formulated an early hypothesis of biological evolution. Putting two and two together, Ionians realized that nature was changing constantly.

Aristotle generally embraced Ionian-style empiricism and acceptance of change when dealing with matters on the Earth, such as biology. When it came to sky phenomena, however, the mystics left their mark on Aristotle. He fell victim to the Pythagorean magical thinking that celestial objects were perfect spheres moving in perfectly circular orbits. He also applied an assertion of Parmenides that nothing ever changed. The resulting Aristotelian view was that stars, Sun, and planets were permanent features with permanent geometric forms, while the Earth was corrupt and therefore imperfect. To explain the dark features on the lunar surface, which conflicted with the idea of perfection, Aristotle suggested that the Moon was contaminated by its proximity to Earth and to the corruption that existed on Earth's surface, namely humans and other forms of life.

SEE ALSO: Quarter-Phase Moon and Heliocentrism (3rd Century BCE), On the Face in the Moon's Orb (1st–2nd Century BCE), The *Almagest* (c. 150), Seeing the First Sliver of a New Moon (11th Century)

Italian Renaissance painter Raphael portrayed many intellectual luminaries from classical antiquity in *The School of Athens* (1511). The two facing figures in the center are Plato (left) and Aristotle (right).

LIBRARY OF ALEXANDRIA

D URING ANTIQUITY, THE MOON DROVE the emergence of astronomy, and astronomy drove the emergence of science as a whole. By the third century BCE, science was concentrated in Alexandria, Egypt. The founder of the city, Alexander of Macedon (356–323 BCE), was a champion for science, as was his most trusted general, Ptolemy I Soter (c. 367–282 BCE), who became ruler of Egypt, the founder of a dynasty that lasted until the famous Cleopatra VII (69–30 BCE). During his 41-year reign, Ptolemy I oversaw construction of an enormous research center, the Museum of Alexandria. Construction continued under Ptolemy's son, Ptolemy II Philadelphus, leading to a facility that included research labs, cages for exotic animals, a medical school with human cadavers, living quarters for visiting researchers, and support for resident scholars who were placed on the royal payroll. Finally, the Museum was linked with a library whose collection began with Aristotle's private book holdings transported from Athens.

Alexandria's library and museum went through ups and downs, and lasted for some six centuries, with the greatest peak of intellectual activity during the reigns of Ptolemy II (reign 285–46 BCE) and Ptolemy III (reign 246–222 BCE). During this early Ptolemaic era, the size of the Library's book collection swelled to an estimated half million to one million scrolls. Such high value was placed on the acquisition of books that the Alexandria police were tasked to search arriving ships for new books, which then were taken to library scribes to be copied on papyrus and made available to scholars.

Scholars at the Library included the geometrist Euclid of Alexandria (mid 300s–285 BCE) and various astronomers, physicists, and mathematicians who were vital to the emergence of lunar science, including Eratosthenes of Cyrene, Archimedes of Syracuse, and Claudius Ptolemy (c. 100–70 CE). The Library also hosted an Ionian researcher, who, by incorporating geometry and precise measurements, would alter the course of astronomy for centuries to come, with the Moon taking on a leading role. His name was Aristarchus of Samos, and we turn to his story next.

SEE ALSO: Beginnings of Nonreligious Astronomy (6th Century BCE), Aristarchus Measures Lunar Diameter and Distance (3rd Century BCE), Quarter-Phase Moon and Heliocentrism (3rd Century BCE), Eratosthenes Calculates Earth's Circumference (3rd Century BCE), *The Sand Reckoner* (3rd Century BCE)

This nineteenth-century engraving shows scholars at work at the Library of Alexandria. The library stored a vast collection of papyrus scrolls, such as those on the shelves in the background of this image.

ad M B per-
pédicularis .
parallela igi-
tur eſt CM ip
ſi LX. eſt au-
tem & SX pa-
rallela ipſi M
R; ac propte-
rea triangu-
lum LX S ſi-
mile eſt trian
gulo M R C.
ergo vt S X
ad MR , ita S
L ad RC. ſed
S X ipſius M
Rminor eſt,
quàm dupla;
quoniã & X
N eſt minor,
quàm dupla
ipſius MO. er
go & SL ip-
ſius CR mi-
nor erit, quã
dupla : &
R multo mi-
nor , quã du-
pla ipſius R
C. ex quibus
ſequitur S C
ipſius CR mi

noré eſſe, quã triplã. habebit igitur R C ad C S maio M

ARISTARCHUS MEASURES LUNAR DIAMETER AND DISTANCE

I N HELLENISTIC TIMES, THE MOON BECAME the key to measurement of its own size and distance to the Earth, as well as the distance between the Earth and Sun, and the relative sizes of all three bodies. Aristarchus of Samos was the central figure in all this work. Samos was an Ionian island, and Aristarchus harbored the Ionian tradition of seeking answers through the observation of nature. But Aristarchus, though born on Samos, was educated in Alexandria. His lifetime overlapped the reigns of the first three Ptolemies. Thus, it is no surprise that Aristarchus remained in the Egyptian metropolis and availed himself of the great Library and Museum, and of the royal funding that was available to support scholars.

Aristarchus was the first person to combine a Greek-style quest for physical models of the universe with Babylonian-style quantitative measurement. From this perspective, the Moon came to center stage through a series of insights. One insight was that by measuring how long it took for the Moon to darken during a lunar eclipse, and then to remain totally eclipsed, Aristarchus would really be measuring the size of the Moon relative to the Earth's shadow.

This taught him that the Earth's diameter was about 3.5 times the lunar diameter. Given that a solar eclipse happened because the lunar disc viewed from Earth was just barely large enough to cover the solar disk, Aristarchus realized that he could scale down the blocking effect. Holding a coin of a certain diameter in front of his face at a certain distance, he could block a full moon, in the same way that the Moon blocked the Sun. Drawing a diagram, he then used simple geometry to estimate the distance to the Moon of about 35 lunar diameters, or 10 Earth diameters. This is about one third of the real distance to the Moon by today's measurements, an impressive achievement for someone whose only instrument consisted of a coin held in front of his face. And yet the genius of Aristarchus shined even brighter in his utilization of the Moon to find the distance from Earth all the way to the Sun.

SEE ALSO: Impact Forms the Aristarchus Crater (450 Million Years Ago), Quarter-Phase Moon and Heliocentrism (3rd Century BCE), Eratosthenes Calculates Earth's Circumference (3rd Century BCE), *The Sand Reckoner* (3rd Century BCE)

By holding a coin in front of his face to cover the full moon, Aristarchus was able to draw a diagram from which this one was ultimately copied. The diagram enabled him to estimate the relative sizes of the Moon and Earth and the distance from Earth to the Moon.

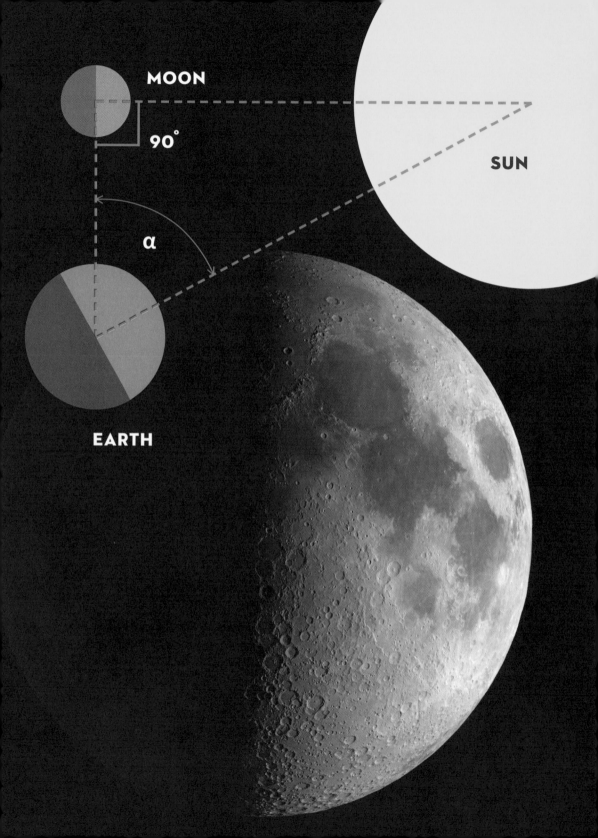

MOON

90°

α

SUN

EARTH

QUARTER-PHASE MOON
AND HELIOCENTRISM

ARISTARCHUS OF SAMOS IS REMEMBERED for producing a cosmological model that featured Earth orbiting the Sun, an achievement that rested on several of his discoveries coming to fruition. His diagram of the relative diameters of the Moon, Earth, and Sun hints at his geometric thinking, but no diagram has been preserved illustrating his greatest insight: recognition that the Earth, Moon, and Sun formed a right triangle when the Moon was precisely in a quarter phase.

With the quarter-phase Moon, Aristarchus set out to measure the angle between two imaginary lines emerging from Earth's center, one pointing toward the Moon, the other toward the Sun. This measurement would dictate the ratio of the distance to the Sun compared to the distance to the Moon. Using crude geometric methods, Aristarchus was able to home in on the ratio, finding that the Sun was between 18 and 20 times the distance to the Moon. Now, the actual distance to the Sun is almost 400 times the distance to the Moon, so Aristarchus was off by a factor of 20. But this was because his crude angle measurement came out to 87 degrees, while the actual angle is 89 degrees, 51 minutes.

His method was therefore sound, as was his reasoning about how to apply his result. If the Sun looked the same size as the Moon, yet was 20 times the distance, then the Sun was almost six times the size of the Earth. To Aristarchus, it seemed absurd that a large Sun should orbit a small Earth, but Aristotle of Stagira had considered and rejected a Sun-centered universe a century earlier, because nobody observed any stellar parallax, an apparent movement of stars that should occur if Earth were traveling through space. For stellar parallax to be unobservable, Aristotle had noted that the stars would have to be unimaginably distant. Aristarchus's answer to Aristotle was that, indeed, the stars really were unimaginably distant. And being so unimaginably distant, Aristarchus proposed, the stars must actually be suns in their own right—suns that probably had planets and life of their own.

SEE ALSO: Impact Forms the Aristarchus Crater (450 Million Years Ago), Aristarchus Measures Lunar Diameter and Distance (3rd Century BCE), Eratosthenes Calculates Earth's Circumference (3rd Century BCE), *The Sand Reckoner* (3rd Century BCE)

When the Moon is in a quarter phase (as shown in the background of this diagram), the Earth, Moon, and Sun create a right triangle. By measuring angle α, Aristarchus could calculate the distance beween the Earth and Sun and the size of the Sun relative to the size of Earth.

ERATOSTHENES CALCULATES EARTH'S CIRCUMFERENCE

N THE CALCULATIONS OF ARISTARCHUS, the distance from Earth to the Moon and from Earth to the Sun were not absolute distances, but proportions. Aristarchus calculated the Earth–Moon distance as being 20 times the radius of Earth, while the Sun's distance from Earth came out to 19 times this distance, or 380 Earth radii. To translate these proportions into absolute distances, thereby revealing the actual size of the Moon and the distance to it, somebody needed to measure the size of the Earth. That man was Eratosthenes of Cyrene, who, like Aristarchus, worked at the Library of Alexandria. Eratosthenes, in fact, was the chief librarian. As such, he spent much time reading. This was a time when the Library, and its associated Museum, were thriving and books were being acquired from all over the known world.

One book in the Library informed Eratosthenes about Syene, a city to the south (now named Aswan), where at noon on the summer solstice the Sun shined directly into a deep well and cast no shadows. Using the shadow length of a tower in Alexandria that same day of the year, Eratosthenes saw a way to compute the curvature of Earth's surface between the two cities. It came to 1/50th of a circle, so Earth's circumference would be fifty times the distance to Syene.

To get the distance, Eratosthenes may have hired surveyors to measure it out, or he could have made an estimate, knowing that camel caravans took roughly 50 days to reach Syene, traveling at a particular rate. The Greeks at this time also possessed a kind of odometer that Archimedes of Syracuse had just invented. One way or another, Eratosthenes learned that the distance between the two cities was just over 5,000 stadia, putting Earth's circumference at 252,000 stadia. (A *stadion* was the length of a sports stadium, of which there were several types.) This puts Eratosthenes's calculation of the circumference at 39,564 kilometers, just a smidgen below Earth's true circumference, if he used the shortest type of stadion as his unit of distance.

SEE ALSO: Eratosthenian Period of Lunar Geological Time (3.2–1.1 Billion Years Ago), Library of Alexandria (Early 3rd Century BCE), Seeing the First Sliver of a New Moon (11th Century)

rd Century BCE

Eratosthenes used this well, located in present-day Aswan, Egypt, to calculate the Earth's circumference.

THE SAND RECKONER

ARCHIMEDES OF SYRACUSE IS CREDITED with approaching a form of calculus, creating numerous mechanical inventions, defining buoyancy and water displacement, calculating the value of pi, and contributions reaching into still more fields. Not surprisingly, his work connected with the study of the Moon.

In his book *The Sand Reckoner*, Archimedes calculated size ratios of the Moon, Earth, and Sun, and used these ratios to estimate that $8x10^{63}$ grains of sand would fill the universe—if Aristarchus were correct that Earth orbited the Sun despite the absence of any measurable stellar parallax (changes in the apparent positions of stars). Archimedes was not embracing Aristarchian cosmology, nor dismissing it. He simply needed a starting value for the size of the universe to make his sand calculation. However, in summing up Aristarchus, Archimedes transmitted the heliocentric hypothesis to future generations, even as Aristarchus's own writings were lost.

Although he studied in Alexandria, Archimedes spent his last few decades in his native Syracuse, a Greek colony under the sway of the Carthaginian Empire. Carthage and Rome were opposing superpowers, and

Archimedes was killed when the Romans held Syracuse under siege in 212 BCE. Along with various heavy-weapons systems, Archimedes was said to have engineered a shoebox-size device that made calculations. Such a device would have operated based on novel technology involving complex systems of small gears that Archimedes's mechanical-engineering protégés continued developing after his death.

Possibly this technology, or components of it, contributed directly to the advent of other calculation devices over the next several decades, including a contraption that sank with a Roman ship in the first century BCE, and that was capable of calculating centuries' worth of motions of the Moon, Sun, and planets. In recent years, the identity and era of the device's creator have been subjects of controversy, but there is reason to suspect that the design depended, at least partly, on Hipparchus of Nicaea, a researcher who lived about a century after Archimedes, and who is remembered as the greatest astronomer of antiquity.

SEE ALSO: A Lunar Facelift (3.9–3.1 Billion Years Ago), Library of Alexandria (Early 3rd Century BCE), Eratosthenes Calculates Earth's Circumference (3rd Century BCE), The Antikythera Mechanism (c. 100 BCE)

A portrait of Archimedes, painted c. 1766 by Italian artist Giuseppe Nogari (1699–1766).

APPLYING MATH
TO THE LUNAR ORBIT

THOUGH LARGELY REJECTED FOR HIS HELIO-centric model, Aristarchus of Samos stands as a revolutionary in science, because he made his discovery based on quantitative measurements of the Moon and Sun, Babylonian style. This approach to astronomy reached new heights, both in Hellenized Mesopotamia and in the heart of the Greek world. In Mesopotamia, Seleucus of Seleucia (c. 190–150 BCE) was first to hypothesize the Moon causing sea tides, but the tides' complexity indicated that Earth must be in motion, so he became one of the few ancient astronomers to adopt Aristarchus's heliocentric model.

Meanwhile, Hipparchus of Nicaea was at work on the island of Rhodes. Remembered for cataloguing positions and brightnesses of more than 800 stars and for discovering precession, a 26,000-year cycle of the wobbling of Earth's spin axis, Hipparchus made extensive use of Babylonian astronomical observations recorded over many centuries made by applying arithmetic and a base-sixty number system. Combining Babylonian mathematics and Greek geometry, Hipparchus was an early developer of trigonometry in the Western world.

Using the Moon's motion as the key to studying orbital motion, Hipparchus repeated the measurements of Aristarchus involving the relative sizes of the Moon and Earth and angles between the Moon, Earth, and Sun. But in 141 BCE, Hipparchus used a solar eclipse to get a much more accurate measure of the Earth–Moon distance. He found it was 429,000 kilometers, just slightly more than what's accepted today as the average distance (384,400 kilometers).

Applying his trigonometry knowledge, Hipparchus also confirmed Babylonian arithmetic-based findings that the Moon's speed changed as it orbited Earth. This implied that for Sun-centered cosmology to be correct, the Moon's orbit must be elliptical, not circular. This was correct, but the Pythagorean-Platonic concept of circular orbits was engrained so deeply into Greek mentality that Hipparchus used his discovery as a rationale for rejecting the cosmology of Aristarchus.

SEE ALSO: Beginnings of Nonreligious Astronomy (6th Century BCE), Aristarchus Measures Lunar Diameter and Distance (3rd Century BCE), Quarter-Phase Moon and Heliocentrism (3rd Century BCE), The Antikythera Mechanism (c. 100 BCE)

This woodcut illustration from the nineteenth century depicts Greek astronomer Hipparchus observing the night sky above Alexandria.

THE ANTIKYTHERA MECHANISM

To account for the Moon's apparent acceleration and deceleration without rejecting circular orbits, Hipparchus embraced two concepts of an earlier astronomer, Apollonius of Perga (c. 240–190 BCE). One concept was that the center of the Moon's circular orbit was not Earth, but a point in space called an eccentric. The other concept was that orbiting objects moved in small orbits called epicycles, superimposed on bigger orbits called deferents. The eccentric was at the center of the deferent, enabling Hipparchus to model the Moon changing speed while going around Earth in circles. Epicycles also accounted for retrograde motion of planets, reversals in the direction of planetary motion with respect to the stars, as observed from Earth.

In 1900 CE, Greek sponge divers pulled an object from a Roman ship that had sunk around 65 BCE near the island Antikythera. Now called the Antikythera mechanism, the object has been subjected to various imaging studies since the mid-twentieth century. Apparently, it utilized at least 37 tiny bronze gears, arranged in complex trains, similar to, but more miniaturized than, clockwork technology that would not reemerge until the high Middle Ages. Yet this device was more than a clock: it was a mechanical computer that coordinated astronomical data, including lunar phases; positions of the Moon, Sun, and planets; and eclipses, accounting for Saros cycles, Meton's 19-year lunar-solar calendar, and various other cycles, across many centuries of time.

Reverse-engineering of gear trains, inscription analysis, and claims by the Roman orator Marcus Cicero (106–43 BCE) that he'd seen two bronze geared mechanisms—one built by Archimedes, the other by Posidonius of Apameia (c. 135–51 BCE), a student of Hipparchus—come into play as historians consider who may have built the Antikythera mechanism. Gear trains reveal what mathematicians call a Fourier series controlling the lunar dial, suggesting a designer who relied on the epicycle-based model that Hipparchus developed in his observatory on the island of Rhodes. Other information extracted from the gear arrangement also points to Rhodes and Hipparchus, or Posidonius, but a dial on the mechanism includes months from the calendar of Corinth, the Greek city that colonized Syracuse, the home of Archimedes.

SEE ALSO: *The Sand Reckoner* (3rd Century BCE), Applying Math to the Lunar Orbit (2nd Century BCE)

A contemporary model of the Antikythera mechanism, the world's first analog computer. It was operated by turning a hand crank located on the side of the device.

ON THE FACE IN THE MOON'S ORB

"The mind is not a vessel to be filled,
but a fire to be lighted."

So wrote the ancient Greek scholar Plutarch (Lucius Mestrius Plutarchus; c. 46–120 CE), as quoted by Apollo 15 commander David Scott in a press conference after returning from the Moon in 1971.

Plutarch's legacy is primarily as a biographer of several leaders of antiquity, but his connection with the Moon extends beyond the citation in the Apollo 15 press conference. For historians of astronomy, it is hard not to take notice of Plutarch's dialogue, *Concerning the Face Which Appears in the Orb of the Moon*, which he included among the 78 essays and transcribed speeches that comprised his work *Moralia*.

Though born four centuries after Aristotle had flourished, Plutarch lived in an age when Aristotle was considered the ultimate authority on the nature of the Earth and heavens. This included the Moon, which Aristotle had considered to be a kind of border area between the heavens, with their never-changing perfect spheres and circular orbits, and the imperfect corruptions of the surface of the Earth. Consequently, the dark regions, to the Aristotelian mindset, were the result of mixing between the perfect and imperfect regions of space.

Aristotle's influence on astronomy would persist for many more centuries, but apparently Plutarch's essay suggests that he took Aristotle's idea with a grain of salt and held a more realistic concept concerning the lunar surface. Essentially, Plutarch's essay describes the lunar dark features as resulting from shadows, due to the presence of chasms, rivers, and other depressions that simply prevent the Sun's light from being reflected: "For it is not possible that a shadow remain upon the surface when the sun casts his light upon all of the moon that is within the compass of our vision" (*Moralia* 5:33–34).

To be sure, Plutarch's thoughts about the Moon were not scientific, as they included ideas that the lunar surface served as a home for the souls of dead people. Nevertheless, he does seem to have considered the dark regions essentially the same as people of his time must have considered the geographical features on Earth.

SEE ALSO: Heavenly Perfection Corrupted (c. 350 BCE), Telescopic Study of the Moon Begins (1609), Learning to Rendezvous and Dock (1965–66), Extended Missions (1971)

A marble bust from ancient Athens of a man believed to be Plutarch.

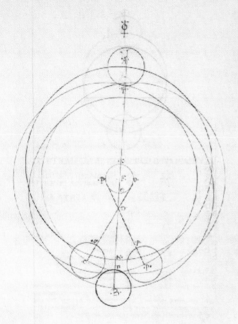

pro+ semidiameter deformeis ct̄. 60

		.60	.22	104	.27.	117	.2
Ac semidiameter epicicli simple. z. f̄ et̄ et̄ addita							
d. b. maxi lon̄ et̄ de elibra	.3.	.0.	6	17.	4.	41.
d. b. minima lonḡ tant̄	.69.		32		129.	24.
d. o. d. p fieb. minui iprei et apo̅no	.33.		32	240	62.	21.
A. q. semidiameter epicicli rota diameter a.b.	.22.	.30	.32	7	83.	53
b. x. maxima lonḡir̄	.26		219	.9.	173.	.28
d. o. d. z̄ mic addit rouch	.23.	.30.	7	69.	130.
	.23.					

Minimam maximamq̄ longitudinem stea secūdā semidiameter nerorum̄ ert̄ ut̄ prī̄po huius libri sinulco d̄ iuo plā̄taū agemur diximus

INCIPIT LIBER · X · DE · MONSTRATIO · MAXIME · LONGITUDINIS · STEALLE · VENERIS · CAPITULUM · I ·

E requisit̄ et̄ secundam porter maximas irsurer distan̄ ria̅ stelle uni̅ luce a fi·e· he tue innuerius lonḡ q̄ equalibus centrum epicicli qinq̄q̄ parte euidē irone lōgītudinem der̄tis equalis fint̄ meem̄ ut̄ iterī irmil̄ uieb̄ suo̅ cepi duas equales hee modo maximā d̄ tem̄tias i̅ hoʒ tieb̄ stelle minduu i̅odices prin̅to int: ipsas demētie termin̅us aut̄ te p̄ q̄irenet̄ lōgītudinem est̄ d̄ uel accrem̄ irtonsi̅ partem q̄i uteǫ distance inuidem̄ irarī ma uel̅ maxima lōgītudine capia debere uel̅ tor̄to minus parmi est̄ magnitudo sua fieb̄ irat̄ irlebbe· 28. ʒ lōgītudine iropilaqum est̄ q̄em̄· i· 31· fimul· o· 26· irngenit̄ autem dt̄at̄ irentirī irete irurge pra· o· 29· Quar̄ irtis ipsiu̅ pro magnitudine ur̄me o· 5· si̅daxit̄ dreduu̅ lıne plen̄ parteb̄ hoʒ est̄ o· 29· lunāt̄ enim̅ mode irurrerirnar lōgītudinī cepi q̄ diameter eru̅ cst̄ o· 14· ʒo· aqonim̄ dur̄ partis fine o· 29· prescrine

DE EPICICLI EIUS MAGNITUDINE

H̄ hoʒ capitulo magnitudinem epicicli nesteri aurtēte sirū

THE *ALMAGEST*

OST OF WHAT IS KNOWN ABOUT Hipparchus's model of lunar and planetary motion comes to us from Claudius Ptolemy. An astronomer at the Library of Alexandria, this Ptolemy was not part of the family of Ptolemies that had ruled Egypt until 30 BCE, but he was Greek, immersed in Greek intellectual traditions. Ptolemy wrote on many topics, but his 13-volume work the *Almagest*, in which he described ideas from Hipparchus and other previous astronomers and weaved in his own material, presented a model of lunar and planetary motion that would dominate astronomy for fourteen centuries.

Along with extending precise star-mapping observations of Hipparchus to include a greater number of stars, Ptolemy expanded on Hipparchus's orbital mechanics. In addition to the eccentrics and epicycles of earlier systems, he introduced equant points. The eccentric, the center of an object's deferent orbit, was located between Earth and the equant, while the epicycle was a smaller orbit superimposed on the deferent. Moving in an epicycle that moved along a deferent, a celestial object would seem to change speed in Earth's skies, but its motion would appear uniform to an observer on the hypothetical equant in space.

The physical corollary to Ptolemaic celestial mathematics was a system of rotating crystalline spheres. Aristotle had conceived this centuries earlier, but now it was condensed and defined mathematically. Ptolemy attempted to account for the Moon's changing speed, and changing distance from Earth, with an inner sphere that carried the Moon without an epicycle. Ptolemy didn't believe that his model of crystalline spheres represented reality, but he simply sought to predict positions of celestial bodies accurately. Mathematically, nothing within the system moved in circles, or with uniform speed. But his model would preserve the illusion of uniform circular motion of celestial bodies through the Middle Ages.

But getting the Moon's motion just right was no easy task. Doing this left Ptolemy with an exaggerated estimate of lunar distance variation. It also would inspire medieval astronomers to start tweaking the Ptolemaic model, and others to overtly critique it.

SEE ALSO: Heavenly Perfection Corrupted (c. 350 BCE), Eastern Astronomers Keep Looking Up (500–800), Shukuk (9th–11th Centuries), The Moon Orbits Alone (1543)

A diagram explaining the movements of celestial bodies from Greek scholar George of Trebizond's commentary on the *Almagest* (c. 1451).

EASTERN ASTRONOMERS
KEEP LOOKING UP

THE TITLE OF PTOLEMY'S MASTERPIECE, *Almagest*, sounds Arabic, because it is an Arabized form of a Greek word. Scholars who translated the 13-volume work placed the Arabic article *Al*, for "the," before the Greek word megiste ("greatest"). The title thus came to mean *the greatest composition*, for it had been the most comprehensive astronomical text to survive from antiquity. Indeed, for centuries after the decline of Greco-Roman civilization, the *Almagest* was the only comprehensive astronomy textbook.

In the *Almagest* celestial system, the Moon was embedded in the innermost of the spheres that rotated around the Earth. This was because the Moon's proximity and swift motion, relative to the other planets, were obvious to the ancient Greeks. Lunar swiftness, plus the Moon's frequent visibility night and day, made the Moon a catalyst for the advancement of astronomy, which in turn led the way for science overall. As Christianity took strong hold of the Eastern Roman Empire, the great library of Alexandria was destroyed and the philosophical schools in Greece were forced to close. Science fell out of favor in the very places where it had led the world.

In Persian lands and beyond, astronomers continued looking up, sometimes making a few fundamental discoveries. Similar to Anaxagoras a millennium earlier, the Indian astronomer Aryabhata (476–550 CE) was teaching around 500 CE that the Moon shined by reflected sunlight, as did the five visible planets. Aryabhata came to this idea through precise astronomical measurements, which also told him that the Earth was spinning. His spinning Earth was at the center of the universe, not orbiting the Sun, but the unusual model would influence another astronomer, five centuries later, to think that the Sun-centered universe might be correct. Meanwhile, Indian astronomers were compiling precise measurements of lunar, solar, and planetary motion into the *Zig al-Sindhind*, a Sanskrit table that would be translated into Arabic in the late 700s CE, from a new city called Baghdad. This would happen under the Abbasid dynasty at the onset of what many historians now call the Arabic Golden Age.

SEE ALSO: The *Almagest* (c. 150), Shukuk (9th–11th Centuries), Seeing the First Sliver of a New Moon (11th Century)

A painting of Indian astronomer Aryabhata. He used astronomical measurements to make many important discoveries, including that reflected sunlight illuminates the Moon.

SHUKUK

HUKUK IN ARABIC MEANS "DOUBTS," A word that several medieval Arabic astronomers used in connection with the system of spheres described in Ptolemy's *Almagest*. Ptolemy's system would hold sway for fourteen centuries, because it predicted celestial positions fairly well. But it just barely held sway, due to improving mechanical devices that didn't magnify celestial bodies, but did track their positions with increasing accuracy. Orbiting Earth in just four weeks, big, and usually visible, the Moon was at center stage, with details of its motion that didn't quite fit the Ptolemaic model.

Setting out not to overturn Ptolemy's model but to correct it, Arabic astronomers, such as al-Battāni (858–929 CE) of Syria, compiled new measurements on lunar and solar motion. Six centuries later, in his rationale for why a heliocentric model must replace the ptolemaic system, Nicolaus Copernicus would cite al-Battāni by name. Along with astronomy, al-Battāni advanced trigonometry, an area of mathematics that Hipparchus had introduced in cruder form. Like Hipparchus, al-Battāni applied trigonometry to his celestial measurements, but his trigonometry was improved with sine and cosine functions replacing old

Greek functions called chords. This helped him compute how distances between the Moon, Sun, and Earth could vary, thereby enabling predictions of whether solar eclipses would be annular (appearing as a ring of sunlight encircling the dark Moon), or complete (the Moon covers the Sun entirely).

As al-Battāni was flourishing in Syria, an institution called Bayt al-Hikma, the "House of Wisdom," was operating in Baghdad, translating ancient Greek works and gathering researchers from multiple fields. Armed with Chinese papermaking technology, the Baghdad scholars could make multiple copies, facilitating a kind of medieval information age. In the late tenth century CE, this information-driven culture produced Ibn al-Haytham (c. 965–1040), an early pioneer in both optical physics and the scientific method, who insisted that Ptolemy's system could not work. Highlighting errors related partly to the system's not accounting for all the details of lunar motion, al-Haytham noted that the rotating spheres carrying the planets would simply collide.

SEE ALSO: Heavenly Perfection Corrupted (c. 350 BCE), The *Almagest* (c. 150), Seeing the First Sliver of a New Moon (11th Century CE), The Moon Orbits Alone (1543)

Ten-dinar Iraqi banknotes commemorate Arab mathematician and scientist Abū Alī al-Hasan ibn al-Hasan ibn al-Haytham, also known as Ibn al-Haytham. Dubbed the "father of optics," he is honored for his achievements in mathematics, physics, and astronomy.

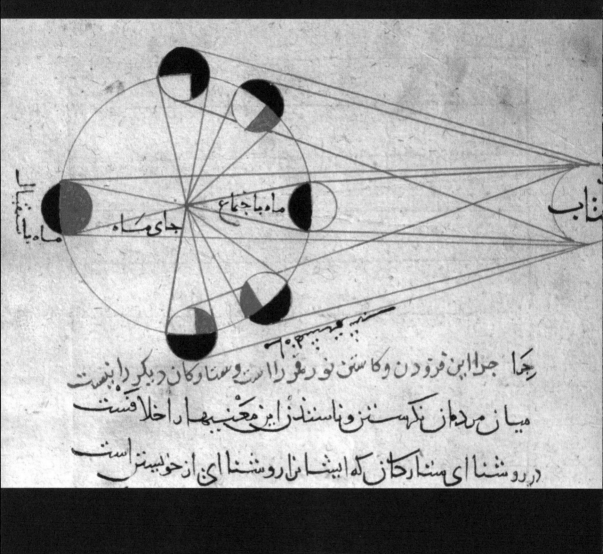

ماه با جماع   جای ماه   ماه یا
آفتاب

رجوع جدا این فرو دن و کاستن نور راا یستن و معناوکان دیگر را بنبست

میان مردمان نگرستن و نا سنندز این معنیها را خلا فست

در روشنای ستار گان که ایشان را روشنای از خویشتر است

SEEING THE FIRST SLIVER
OF A NEW MOON

PRECISE TIMING OF LUNAR-PHASE CYCLES was vital for Muslims, Jews, and others within Islamic caliphates. Where Islam followed a purely lunar calendar, Judaism depended on a Metonic-style lunisolar calendar developed from the Babylonian calendar by astronomer Samuel of Nehardea (c. 177–257 CE). In both religions, appearance of a thin crescent trailing the sunset signified the beginning of the new month, with various implications on timing of rituals. In the fourth century, Jewish leaders sought to fix the timing of future holidays based on astronomical calculations, although for centuries they continued having witnesses observe each new moon with their eyes, as did Muslims.

It was not always so easy to see such a thin crescent. Moreover, witness accounts of the first sliver did not always correspond to the calculated days for starting the new month. This was due partly to the complexity of lunar motion; but in the eleventh century, the central Asian astronomer al-Biruni (973–c. 1052 CE) made headway toward a solution with a tubular device that looked almost like a telescope. It had no lenses or mirrors for magnification; but in keeping the central view dark, it made the thin crescent of a new moon slightly easier to spot.

A first-rate polymath, al-Biruni's legacy extends beyond the sketches of lunar phases created with his tube. Like Eratosthenes, al-Biruni calculated Earth's circumference accurately, leading him to posit the presence of a continent west of Africa and east of China. Along with a host of sciences, he studied comparative religion, linguistics, and Indology. Expertise in the latter topics followed from Sultan Mahmud of Ghazni bringing al-Biruni to northern India as an adviser during a military campaign. Already a critic of Aristotle and growing skeptical of Ptolemy, al-Biruni now met Indian astronomers. They introduced him to the cosmology of Aryabhata, which featured Earth at the center of the universe, but spinning. Considering Aryabhata, plus the astronomy of the ancient Greeks, al-Biruni wrote that whether Earth was indeed stationary, or whether it was spinning and orbiting the Sun as the Moon orbits Earth (as Aristarchus had proposed), depended on one's perspective.

SEE ALSO: Heavenly Perfection Corrupted (c. 350 BCE), Eratosthenes Calculates Earth's Circumference (3rd Century BCE), The *Almagest* (c. 150), Eastern Astronomers Keep Looking Up (500–800), Shukuk (9th–11th Centuries), The Moon Orbits Alone (1543)

A sketch by Abū Rayhān Muhammad ibn Ahmad al-Bīrūnī showing the different phases of the Moon. al-Bīrūnī made his observations using an early precursor to the telescope that he developed himself.

A NEW MODEL
FOR LUNAR MOTION

WHEREAS AL-BIRUNI QUESTIONED the entire geocentric framework, other astronomers adjusted the Ptolemaic model to account for inconsistencies with lunar motion and eliminate other problems. In the thirteenth century CE, the Persian astronomer Nasir al-Din al-Tūsi (1201–1274) began this task with a new model for lunar motion. Yet al-Tūsi and al-Biruni were kindred spirits. From their writings, it's clear that they dismissed astrology as superstition, despite serving as official astrologers for powerful men. Al-Biruni's paycheck was for supplying Sultan Mahmud of Ghazni with battle horoscopes; and when the Mongols invaded, al-Tūsi convinced the Hulagu Khan that he could help him win more battles, if only he had a new observatory with advanced instruments. This ploy gave rise to the Maragheh observatory, where astronomers from lands as distant as Iberia and China gathered to do research. Using al-Tūsi's state-of-the-art 10-meter armillary arm, essentially a giant protractor, they plotted positions of the Moon, planets, and stars with unprecedented precision. This led them to propose new systems for celestial motion that made Ptolemy's model more palatable. These systems included the "Tusi couple," wherein planets moved on circles rotating within bigger circles, and also a new mathematical model of lunar motion. Geometrically, the Tusi couple was different from epicycles, and it enabled removal of the equant points that Ptolemy had used as a mathematical trick to say that celestial objects moved with uniform motion, despite the fact that they did not move that way as viewed from Earth.

This reformation of the Ptolemaic system would help sustain geocentric cosmology for another two centuries. During that time, one astronomer in Provence, Rabbi Levi ben Gershon (1288–1344), was using planets and the Moon to estimate the minimum distance to the stars. Being Jewish, he was well versed in Arabic. This kept him abreast of Arabic researchers, who still led the world in astronomy. And while he's often excluded from the story of the demise of Ptolemy's model, his estimate of stellar distance effectively confirmed the heliocentric model of Aristarchus.

SEE ALSO: Lunar Brightness to Estimate Stellar Distances (14th Century), Adjusting Lunar-Distance Variation (14th Century), The Moon Orbits Alone (1543)

Muhammad ibn Muhammad ibn al-Hasan al-Tūsi, also known as Nasir al-Din al-Tūsi, sits at his writing desk in this thirteenth-century illustration. He lectures his pupils at the Maragheh observatory, located in present-day northern Iran.

LUNAR BRIGHTNESS
TO ESTIMATE STELLAR DISTANCES

O N THE MOON SITS THE CRATER RABBI Levi, named for Levi ben Gershon, a Provençal Jewish astronomer known by the Hebrew acronym "RaLBaG," and by his Latinized name, Gersonides. A polymath-philosopher whose Aristotelian empiricism led him to natural explanations for "miraculous" biblical events, Gersonides might have irked Christian authorities. But the papacy in Avignon supported his astronomy with enthusiasm, possibly because he wrote his biblical commentaries in Hebrew, which only Jews understood.

The late physicist Yuval Ne'eman (1925–2006) highlighted Gersonides as the only astronomer, prior to the nineteenth century, to obtain a realistic estimate of stellar distances. Measuring the actual distance to a particular star would not happen until 1838. In that year, Friedrich Wilhelm Bessel (1784–1846) would record changes in stellar parallax that enabled him to compute 11.4 light-years as the distance to the star system 61 Cygni. Five hundred years earlier, technology was inadequate for detecting stellar parallax. Titanic armillary arms, like what al-Tūsi had at the Maragheh observatory, could possibly have enabled an astronomer with excellent eyesight to detect parallax of the Alpha Centauri system, the nearest to us, located "only" 4.3 light-years away, but such an instrument would have had to be located in Earth's southern hemisphere.

Consequently, there was no evidence that Earth was moving; and yet, like Ibn al-Haytham and al-Biruni, Gersonides was suspicious of the Ptolemaic model. He realized that by recording changes in lunar brightness and size and comparing his measurements with available figures for distances to the Moon over the course of lunar cycles, he could calibrate lunar distance with brightness. Applying similar methods to look for changes in the brightness of Mars might reveal cycles supporting, or disproving, Ptolemy's epicycles. Finally, Ne'eman suspected that Gersonides had combined his findings on lunar brightness with available figures on lunar and planetary distances to produce his estimate of the minimum possible distances to typical stars. Whatever the details of his method, Gersonides ended up estimating that the stars of the Big Dipper must be at minimum 10 to 100 light years away.

SEE ALSO: Shukuk (9th–11th Centuries), Seeing the First Sliver of a New Moon (11th Century), A New Model for Lunar Motion (13th Century), Adjusting Lunar-Distance Variation (14th Century)

An illustration of a man using a Jacob's staff, invented by French–Jewish astronomer Gersonides. Consisting of a pole with markings measuring length, this tool was used in the fourteenth century to determine astronomical distances.

وهذه صورة مدار القمر هو مدار الأكبر المتحرك حسه مصور على السطح الخارج لمراكز
وحساب التعاديل على أنا صورنا طلها مع مدة الدوريف فأما أماكن واللامكان هي الخارج والمعدل
والمعدل المار وضعها على النار وسطى طلها من النار أنصاف الأربع عكفكف
لسهول الصورة كما نور

مركز العالم

وسطها رجه أن وحركة وسط القمر في عشرين سنة فارسيه ب مرطح حا
دى سنه واحده حط ط نب كط كد لوح ملتين يوما اه مرا لوح د وفي يوم بليله
مح دلم الحنب لوح بور فى ساحه مستويه بب نو كزا لجمالا وحنبنان
حاصه القمر للتاريخ المذكور دح لـ كرد وحركتها في عشرين سنه فارسيه باد كبـا
دى سنه واحده سع حور و فى شهر النوح ب دن بعد الـ خ ح ح ه نور سلعه هل
لامه وانبتنا وسط المحبور زهر للتاريخ المذكور ده رله وحركة الحلا الحلاق الوالي ز علا مريل
ن

ADJUSTING LUNAR-DISTANCE VARIATION

WHEN THE EUROPEAN RENAISSANCE started in fourteenth-century Florence, Arabic astronomy still thrived to the east, and the Ptolemaic model for lunar and planetary motion still held a steady sway. But astronomers had reformed the model substantially since Ptolemy had formulated it twelve centuries earlier. During the intervening time, the Moon had orbited Earth more than 15,000 times, with generations of astronomers watching not just its changing phases, but also changes in brightness and in the fraction of arc taken up by the diameter of the lunar disc. These features, plus the fact that the Moon's waxing and waning did not happen at a constant rate, told astronomers that the Moon's speed was always changing, as was its distance from Earth.

To account for these lunar factors and other problems with the model presented in Ptolemy's *Almagest*, the adjusted model for motion of the Moon, planets, and Sun around Earth was, by the fourteenth century, a rather complex network of whirligigs. And yet, even the most novel adjustment—the Tusi couple, added during the previous century—did not eliminate the problem that bothered people

the most, namely eccentricity: the mathematics did not actually leave the Earth physically at the exact center of anything. The adjusted system also did not account for some aspects of the motion of the planet Mercury.

To confront these problems, the Syrian astronomer Ibn al-Shatir (1304–75) added some extra epicycles (smaller spheres that rotated while moving along the circumference of larger spheres), while maintaining al-Tūsi's mathematics. He also adjusted the variation in the Moon's distance from Earth so that it fit with observations better than Ptolemy's lunar variation, which was highly exaggerated. The resulting system accounted better for lunar and planetary motion, including Mercury's. It kept Earth near the center of the universe too, and yet, mathematically, al-Shatir's reformed system would fall right in place with the next big milestone proposal for celestial motion. That next proposal would come 150 years later from a Polish mathematician-astronomer by the name of Nicolaus Copernicus.

SEE ALSO: Shukuk (9th–11th Centuries), Seeing the First Sliver of a New Moon (11th Century), A New Model for Lunar Motion (13th Century)

A diagram by Ibn al-Shatir of the Moon's orbit around Earth. His model for planetary and lunar motion resolved problems with earlier ones, including those from al-Tūsi and Ptolemy.

THE MOON ORBITS ALONE

HISTORIANS HAVE TWO IDEAS TO EXPLAIN what led Nicolaus Copernicus to propose a heliocentric universe. One is that Copernicus was not satisfied with how the reformed Ptolemaic system provided only the illusion of uniform circular motion for the Moon, planets, and Sun. The other is that, in doing the math, Copernicus simply took things to the next logical step from where al-Tūsi and al-Shatir had advanced things.

Support for the latter idea comes from analysis of Copernicus's famous work *De Revolutionibus Orbium Coelestium* (*On the Revolutions of Heavenly Spheres*), which contains much more than just his famous diagram of Earth and other planets moving in circular paths around the Sun, with only the Moon orbiting around the Earth (in pre-Copernican models, the Moon, planets, and Sun all moved around the Earth). More telling is how his work reproduced the mathematics of the Tusi couple and included diagrams derived from drawings by various Arabic astronomers. He mentioned by name the ninth-century astronomer al-Battani, whose astronomical tables he used and supplemented with his own data. Copernicus also preserved mathematics that al-Shatir had added on top

of al-Tūsi's, and described the Moon orbiting Earth with changing distances and speeds that matched al-Shatir's figures very closely.

These realizations have led the British physicist, humanist, and author Jim Al-Khalili to describe Copernicus as the culminating astronomer of the Arabic period, rather than the first astronomer of the phase of science that would place Europe on center stage. As for why such a development would happen in Europe, technological change was at play. The printing press meant that Copernicus didn't need a Library of Alexandria, or a House of Wisdom, but could afford his own book collection. But while his new model did a good job of explaining planetary retrograde, Copernicus could not explain the variations in lunar brightness, and in the speed of the Moon and planets, any better than al-Shatir's model could. Always thinking that he might improve his model, Copernicus waited to publish *De Revolutionibus* until 1543, the year of his death.

SEE ALSO: Copernican Period of Lunar Geological Time Begins (1.1 Billion Years Ago), Shukuk (9th–11th Centuries), A New Model for Lunar Motion (13th Century), Adjusting Lunar-Distance Variation (14th Century)

An illustration of Copernicus's heliocentric model, which contends that Earth orbits the Sun and only the Moon orbited Earth. Copernicus was influenced by the findings and calculations of Arabic astronomers who came before him as he developed his model.

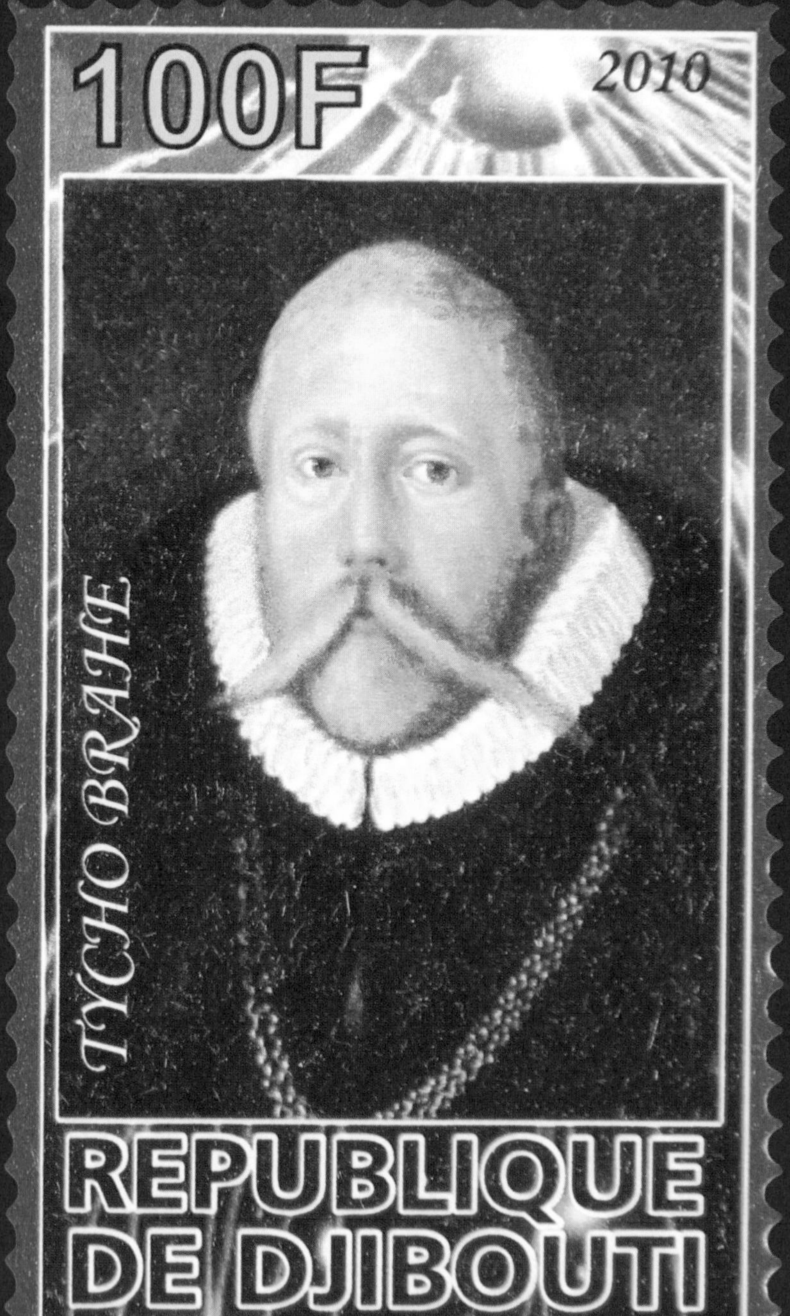
100F
2010
TYCHO BRAHE
REPUBLIQUE
DE DJIBOUTI

MOON AND SUN ORBIT EARTH

THE NOTORIOUS CATHOLIC BANNING OF *De Revolutionibus* did not happen until six decades after its publication; but thanks to the printing press, hundreds of copies of *De Revolutionibus* did spread through Europe, making their way into the hands of astronomers on both sides of the Catholic–Protestant dispute that consumed Europe. One such astronomer was Tycho Brahe (1546–1601), a flamboyant, wealthy Dane who wore a metal prosthesis for a nose that he'd lost at age twenty, sword-fighting with a third cousin over who was the better mathematician.

Eponym of a prominent crater in the southwest of the lunar nearside, Tycho became a celebrity in 1572 when he noticed a star in the constellation Cassiopeia that should not have been there. Tycho easily verified that the new star—which for a few days was brighter than Venus—lacked parallax, so it was not an atmospheric phenomenon. He'd actually seen a supernova, a star in the midst of an explosive death that previously had not been visible because of its distance from Earth. To keep the talented astronomer in Denmark, King Frederick II gave Tycho an island and funds to build Uraniborg, an observatory with state-of-the-art instruments, such as the Tychonian quadrant that could enable precise measurements of the positions of the Moon, Sun, planets, and stars. Disagreements with the next Danish king led Tycho to move to Prague, where the Holy Roman Emperor, Rudolph II, built him a new observatory.

Though intrigued by Copernicus, Tycho ruled out a moving Earth, because he could not detect stellar parallax, despite his instruments and astronomical skill. Instead, he proposed that only the Moon and Sun orbited Earth, while the planets orbited the orbiting Sun. His observational data fit his hypothesis well for Jupiter and Saturn, and adequately for Venus. Mars was problematic, however, so he needed a mathematician even better than himself. Clearly, his nose-chopping cousin was not an appropriate candidate, so he invited a younger man, a German by the name of Johannes Kepler.

SEE ALSO: Copernican Period of Lunar Geological Time Begins (1.1 Billion Years Ago), A Dream of a Lunar Voyage (1581), Telescopic Study of the Moon Begins (1609)

A portrait of Tycho Brahe on a 2010 commemorative stamp from Djibouti.

A DREAM OF A LUNAR VOYAGE

AT AGE NINE, JOHANNES KEPLER (1571–1630) saw a lunar eclipse. Much later, he would write a science fiction novel, *Somnium*, about a scientist whose witch mother helps him travel to the Moon in a dream. The story describes how Earth might look from the Moon, foreshadowing a famous photograph that the astronauts of Apollo 8 would take centuries later. The protagonist also meets Tycho Brahe, a person from Kepler's real life. Sadly, the witch mother would become a self-fulfilling prophecy in 1617, when Kepler's mother, Katharina (1546–1622), would be arrested for witchcraft. Only in 1621 would she be released, thanks to her son's legal skills.

Kepler was educated on a scholarship and expected to enter the Lutheran clergy, but his theological ideas crossed between different Protestant sects. This barred him from high-paying university positions, but he accepted a job teaching math and astronomy at a Protestant school in Graz, Austria. There, despite a plethora of health and financial troubles in his family, he published a book with his mathematical perspective on the Copernican system, leading to correspondence with Tycho Brahe.

In 1600, militant counter-Reformation activity in Graz led Kepler to accept an invitation from Tycho to work in Prague on the problem of Mars not fitting his model of the universe. Soon after Kepler's arrival, Tycho was drinking heavily as a party guest of the emperor. Thinking it rude to leave the emperor's table, Tycho held his urine to the point of damaging his bladder, and he died eleven days later. In his notebooks were numerous observational data, which Kepler found to be consistent with a heliocentric universe, but only if Copernican circular orbits were replaced with a different shape.

That shape was an ellipse, a squashed circle with two foci, one corresponding to the position of the Sun. This was Kepler's first law of planetary motion, which he published in 1609, along with another law stating that a line between a planet and the Sun sweeps across equal areas in equal times. Mathematically, Kepler's findings made a modified Copernican system possible; but as for proving Copernicus, that would require a technology that was just in its infancy.

SEE ALSO: Copernican Period of Lunar Geological Time Begins (1.1 Billion Years Ago), Telescopic Study of the Moon Begins (1609), Advancing Telescopes Eye the Moon More Closely (17th Century), The Moon Inspires Isaac Newton (Late 17th Century)

Johannes Kepler holds court with Emperor Rudolf II, who appointed him imperial mathematician in 1601 after the sudden death of fellow astronomer and mathematician Tycho Brahe.

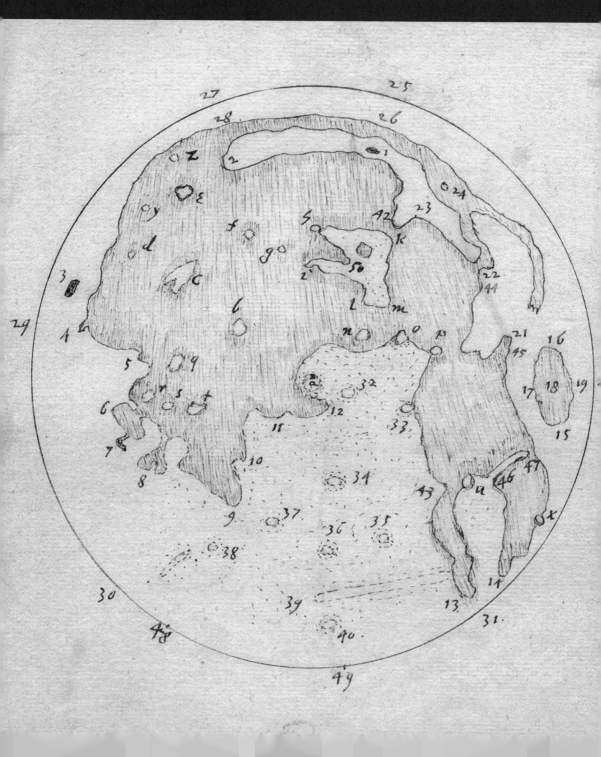

TELESCOPIC STUDY
OF THE MOON BEGINS

N 1971, APOLLO 15 COMMANDER DAVID SCOTT dropped a hammer and feather on the Moon to acknowledge Galileo Galilei (1564–1642) as one enabler of the Space Age. Galileo had performed similar experiments that launched experimental physics, and his contribution to lunar exploration would not end there.

By late 1609, Galileo was pointing telescopes at the Moon, though he wasn't the first. Thomas Harriot (1560–1621) had sketched lunar maps months earlier using a Dutch telescope. Galileo built his own telescopes, achieving a magnification of about nine times before initiating lunar study. Near the terminator that divided lunar day and night, he saw shadows from mountains, elongated because sunlight struck this region at a sharp angle. A mountainous surface implied an Earth-like world, and when Galileo saw moons orbiting Jupiter, Earth's Moon was demoted further. None of this bothered Galileo, who favored Copernicanism. Teaching in Padua, under Venetian sway, he was safe discussing new ideas, so long as he didn't go further than Copernicus. In 1600, the Roman Inquisition had sent Giordano Bruno (1548–1600) to a torturous death for holding unorthodox ideas, including that stars were suns with worlds of their own.

Galileo's collision course with religious authorities began when he relocated to Florence and turned his telescope on the planet Venus. Relocation gave him a position at the University of Pisa, his alma mater, but now with more prestige and no teaching responsibilities. He also gained patronage of the powerful Medici family. The Church could not protest the work of a Medici courtier of exponentiating fame, but observation showed Venus phasing between crescent and circular shapes, like the Moon, except that the crescent Venus looked bigger, and then shrunk as it shifted to a circle. This proved that Venus orbited the Sun. Galileo reasoned that the other planets, and Earth, must be doing the same, as Copernicus had posited. Trouble lay ahead, though, since Venusian phasing still allowed for Tycho Brahe's idea of the planets orbiting the Sun around a stationary Earth. Moreover, politics between the papacy and the Medici family were complex.

SEE ALSO: Copernican Period of Lunar Geological Time Begins (1.1 Billion Years Ago), Quarter-Phase Moon and Heliocentrism (3rd Century BCE), A Dream of a Lunar Voyage (1581), Advancing Telescopes Eye the Moon More Closely (17th Century), Extended Missions (1971)

Thomas Harriot created some of the first lunar maps. This drawing (c. 1609) depicts the Moon's maria and craters, which Harriot observed using a telescope.

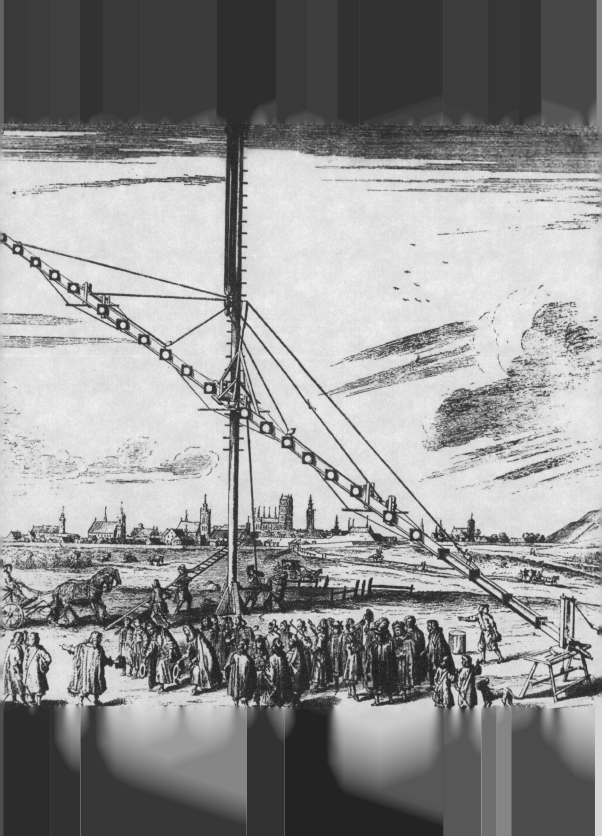

ADVANCING TELESCOPES EYE THE MOON MORE CLOSELY

G ALILEO AND HARRIOT BOTH COMMUNI-cated with Johannes Kepler, who studied optics. Kepler published his first two laws in 1609, the same time that Harriot and Galileo were observing the Moon. Galileo's *Sidereus Nuncius* (often translated as *Starry Messenger*), published in 1610, inspired Kepler to design new telescopes, featuring a convex eyepiece in place of the concave eyepiece of Galilean telescopes. This widened the field of view and reduced eyestrain.

The increasing popularity of telescopes led to more observations, more hypotheses, and more disagreements between astronomers. Disagreeing with Galileo, astronomers found their arguments rebutted to shreds. Galileo made bitter enemies, which ultimately led to agreement that he should present a balanced view of the geocentrism–heliocentrism debate in a paper. What he wrote was an illustrated dialog between fictional characters, one being a wise, witty Copernican, like himself. The geocentric character, named Simplicio, wore a papal hat and recited the pope's favorite talking point. This triggered a series of events, culminating with Galileo's being forced to recant his support for heliocentrism in 1633. He spent the rest of his life under house arrest.

Meanwhile, telescopes grew longer until Johannes Hevelius (1611–87) built one 151 feet long (46 meters). It was too unstable to be practical, but with more modest scopes Hevelius obtained four years' worth of lunar charts that he published in his *Selenographia* atlas in 1647. Using similar telescopes, Michael Florent van Langren (1598–1675) and Francesco Maria Grimaldi (1618–63) also drew detailed lunar maps. Grimaldi's maps were included in *New Almagest*, written in 1651 by Giovanni Battista Riccioli (1598–1671), creator of the system for naming lunar structures. The terra and maria, and the naming of craters for those ancient scholars all come from Riccioli.

Another optical innovation came from Dutch astronomer Christiaan Huygens (1629–95), but telescopes using lenses entailed several problems, one being that lenses separate light into colors. Understanding the light color-separating effect of glass better than anyone else, one English researcher, Isaac Newton, began using curved mirrors as magnifying elements instead of lenses.

SEE ALSO: A Dream of a Lunar Voyage (1581), Telescopic Study of the Moon Begins (1609), The Moon Inspires Isaac Newton (Late 17th Century)

An engraving showing one of Johannes Hevelius's telescopes. Its lenses were held in place by a system of ropes and pulleys that required constant adjustments. The device proved to be impractical for regular use.

THE MOON INSPIRES
ISAAC NEWTON

WHAT PROPELS THE MOON AROUND the Earth, and could it be the same phenomenon that causes objects to fall? This question led Sir Isaac Newton (1642–1727) to explain motion, mathematically, in terms of a universal force. Earlier researchers had discovered elements of Newtonian physics, beginning with Galileo. Straddling the lifetimes of Galileo and Newton was the French mathematician-philosopher René Descartes (1596–1650), creator of analytic geometry. In 1644, Descartes had laid out three laws roughly anticipating the three laws of motion that Newton set down.

Newton's gravitation came to life in the 1670s and '80s. During this period, scholars recognized Newton for experiments on the nature of light and color, published in 1672. They also knew of Newton's achievements in mathematics that had earned him the prestigious Lucasian professorship—the same post at the University of Cambridge that the late Stephen Hawking (1942–2018) would hold. But Newton was initially quiet and private about gravitational theory, writing publicly instead about theology and dabbling in alchemy.

Back in 1619, Kepler had published a third law of planetary motion, stating that the square of a planet's orbital time around the Sun is proportional to the cube of the orbit's semi-major axis (half the long diameter of the ellipse). This expanded on Kepler's other two laws, but Kepler had worked empirically from Tycho Brahe's astronomical observations. Newton, in contrast, demonstrated mathematically the presence of an attraction between any two bodies, a force that is proportional to the product of the masses of the bodies and inversely proportional to the square of the distance between them.

Only after astronomer Edmund Halley (1656–1742) encouraged him did Newton publish his gravity idea in *Philosophiæ Naturalis Principia Mathematica* (*Mathematical Principles of Natural Philosophy*). Printed in 1687, *Principia* accounted for lunar motion and all motion on and off Earth, and predicted Kepler's laws. To make it work, Newton had invented calculus, which together with universal gravitation would initiate modern science and enable the Industrial Revolution and the technology that would take humans to the Moon.

SEE ALSO: A Dream of a Lunar Voyage (1581), Improving Instruments Advance Lunar Astronomy (18th Century), The Moon Proves General Relativity (1914–22)

A portrait of Sir Issac Newton by English painter Sir James Thornill (1675–1734), completed in the early eighteenth century.

IMPROVING INSTRUMENTS
ADVANCE LUNAR ASTRONOMY

THE EIGHTEENTH CENTURY SAW RAPID advances in precision instrumentation, leading to a flowering of astronomical studies. These ranged from the mapping of Earth's magnetic field to the discovery of the planet Uranus, but astronomers also observed the Moon and considered lunar interaction with other astronomical phenomena.

The astronomical giant at the dawn of the eighteenth century was Edmund Halley, Astronomer Royal of Britain for more than two decades. Halley is remembered best for realizing and demonstrating the periodicity of comets; but two decades earlier, he had shifted the course of science by persuading Isaac Newton to write *Principia*, and by paying for its printing. In mapping the geomagnetic field in 1702, Halley resolved a vexing navigational issue, created the science of geophysics, and set the stage for studies on space radiation of major import to NASA's Apollo lunar missions, and for our current communication systems. Halley also realized, and in 1739 confirmed, the secular acceleration of the Moon, the small change in lunar speed due to tidal forces. Immanuel Kant (1724–1804) may have been the first to propose that, through lunar tides producing an ocean bulge ahead of the Moon's pathway, this process also slowed Earth's rotation.

Halley also saw potential in astronomer James Bradley (1693–1762). In 1727, Bradley discovered an aberration effect that astronomers would have to work into studies involving star positions. In 1748 he discovered Earth nutation—a wobble of Earth's rotational axis that he attributed to lunar motion, and that complicated the precession wobble that Hipparchus had discovered centuries earlier. In mid-century, the German astronomer Johann Tobias Mayer (1723–62) was applying new cartographic techniques to map the Moon in unprecedented detail; and, by analyzing how the Moon blocked other bodies, the Croatian polymath-astronomer Roger Joseph Boscovich (1711–87) was proving that the Moon lacked an atmosphere. Decades later, William Herschel (1738–1822), the discoverer of Uranus, reported seeing three red luminescent spots on the lunar surface near the Aristarchus crater, and suggested that they might be active volcanoes.

SEE ALSO: The Moon Inspires Isaac Newton (Late 17th Century), Scientists Consider Lunar Origins (1873–1909), A New Discovery and a New Agency (1958–59)

An illustration of Edmond Halley (left) and Sir Isaac Newton (right) in the middle of a lively discussion about planetary motion.

A LUNAR SOCIETY IN BIRMINGHAM

STEAM POWER, CANALS, MINERALOGY, astronomy, new drugs, the age of the Earth, and gas chemistry—these topics were on the agenda for a group of manufacturers and intellectuals that met monthly in England's midlands, initially at the Birmingham home of industrialist Matthew Boulton (1728–1809). Around 1760, Boulton and Erasmus Darwin (1731–1802) founded an informal think tank that became the Lunar Society of Birmingham. A physician and botanist, Darwin was the author of an evolutionary concept anticipating the theory of natural selection that his grandson, Charles Darwin (1809–82), would develop a century later.

Other "lunartics" included James Watt (1736–1819), whose steam-engine innovations Boulton supported with financing and metal parts, and Charles Darwin's other grandfather, Josiah Wedgwood (1730–95) of pottery fame. Eventually, Joseph Priestley (1733–1804), the chemist who discovered oxygen in air and released from plants, would join too.

The group is remembered as the Lunar Society, because meetings were held near the time of a full moon, so attendees would have light to return home safely. There were many other lunar societies, but the Birmingham society stands out as a group of science and technology promoters that exchanged ideas, and networked with experts outside England, including Caroline Herschel (1750–1848), astronomer and sister of William Herschel.

Outside the group leaders, meetings often included various others, such as William Small (1734–75), a physician who pioneered the drug digitalis. Benjamin Franklin (1706–90) would journey to the meetings from London, where he lived just prior to the American Revolution. Their ranks included deists and Unitarians. Many, especially Wedgwood, Priestley, and Darwin, campaigned for the abolition of slavery.

Meetings would move from dinner to discussions, demonstrations, and experiments. Along with pragmatic manufacturing plans that may have hastened the onset of the Industrial Revolution, the group explored more exotic topics, such as Darwin's proposal to divert the world's navies away from war to the task of hauling icebergs from polar to equatorial regions—to cool the tropics while making England warmer.

SEE ALSO: Improving Instruments Advance Lunar Astronomy (18th Century), Scientists Consider Lunar Origins (1873–1909)

The Soho House was the regular meeting place of the Lunar Society of Birmingham and home of entrepreneur and prominent member Matthew Boulton.

ANOTHER DOCTOR
TURNS HIS EYES TO THE MOON

EVEN BY THE NINETEENTH CENTURY, AN individual could engage in astronomy alongside other careers. Long gone were the polymaths, like Aristotle and al-Biruni, but astronomers as recently as the seventeenth century had often shifted between topics, or gravitated to astronomy after trying something else. Johannes Kepler studied medicine and law before mathematical astronomy, but he had also applied his legal skills to defend his mother against witchcraft charges. Copernicus and Galileo had studied medicine as well.

So it was not so strange when Franz von Gruithuisen (1774–1852), a Bavarian physician and medical instructor who tested novel treatments for kidney stones, was appointed an astronomy professor at the University of Munich in 1826. Nor was it surprising that von Gruithuisen suspected that the Moon might support life forms, as astronomers were seeing the Moon as an Earthlike world. What did make von Gruithuisen an outlier, however, was a report that he published in 1824, that he'd found evidence for lunar life, indeed for a civilization of intelligent beings. Observing the lunar surface, he had spotted and sketched fishbone patterns of ridges that he interpreted as city streets and buildings.

Von Gruithuisen gave his lunar city a name—*Wallwerk*, but he had made his observations with only a small refracting telescope. Armed with more powerful equipment, other astronomers quickly demonstrated that von Gruithuisen's imagination had gotten the better of him. That same vivid imagination put von Gruithuisen generations ahead, however, when he proposed in 1829 that lunar craters had been formed by meteorite impacts. Ironically, one early champion of the alternative hypothesis, that lunar craters were actually volcanic calderas, also had two careers. This was James Nasmyth (1808–90), a Scottish engineer who had shifted into lunar astronomy after getting wealthy from inventing the steam hammer. Astronomy melded naturally with engineering, but lunar science also was attracting people from the fantasy realm, beginning with Jules Verne (1828–1905), whose novel *De la terre à la lune* (*From the Earth to the Moon*) imagined three future humans traveling to the Moon in a capsule, launched from Florida.

SEE ALSO: Seeing the First Sliver of a New Moon (11th century), A Dream of a Lunar Voyage (1581), New Understanding of Craters (1948–60)

This illustration, taken from a page in the first edition of *From the Earth to the Moon* (1865) by Jules Verne, shows his characters experiencing weightlessness on the way to the Moon.

И ТЕМ НАГРАЖДЕНЫ УСИЛЬ
...БОРОВ БЕСПРАВИЕ И ТЬМ'
...ОВАЛИ ПЛАМЕННЫЕ КРЫЛЬ.
...ОЕЙ
...СТРАНЕ

И ВЕКУ СВОЕМУ!

...ЗНАМЕНОВАНИЕ ВЫ...

ЦИОЛКОВСКИЙ

ОСНОВОПОЛОЖНИК
КОСМОНАВТИКИ

VERNE INSPIRES THE FATHER
OF ASTRONAUTICS

A LMOST DEAF FROM THE SCARLET FEVER that had struck him at age ten, Konstantin Eduardovich Tsiolkovsky (1857–1935) could not attend school in Russia, so he studied at home, where he also consumed novel after novel of Jules Verne's *Voyages Extraordinaires* series. In particular, he liked *De la terre à la lune*, a story about three people traveling in a capsule that is launched to the Moon by an enormous cannon. Verne's formal training was in law and the arts, but he used technical detail: calculations of power, acceleration, and distance, for instance. This put Verne far ahead of other fiction writers of the era, but his stories also left blatant errors of physics and physiology. To achieve the high elliptical trajectory needed to reach the Moon, for instance, a craft requires a velocity of almost 11 kilometers (7 miles) per second. But acceleration to this velocity through Verne's "Columbiad Space Gun" would have been lethal to his fictional astronauts.

Educating himself in the basics of Newtonian physics, a teenage Tsiolkovsky wondered if there might be a way to accelerate humans more gradually, but over periods of time long enough to achieve orbital velocity, or even higher velocities to reach the Moon and planets. Showing an aptitude for physics, he attended open lectures, aided by a long earhorn, and studied at the Moscow library, where he met the philosopher, Nikolai Fyodorov (1829–1903). An early transhumanist, as he proposed applying technology to achieve radical life extension, Fyodorov encouraged Tsiolkovsky, who eventually passed an exam that allowed him to teach at a school near Moscow.

While teaching school, Tsiolkovsky published numerous technical articles, similar to Johannes Kepler during his teaching days, but Tsiolkovsky also drafted plans for flying machines that ultimately got him government support to build a wind tunnel. His growing aeronautical authority drew attention to his paper, *Exploration of Outer Space by Means of Rocket Devices*. Published in 1903, this treatise laid out the basics of theoretical astronautics, and may have influenced Sergey Korolyov (1906–66) whose rocket and spacecraft designs for the Soviet Union would launch the Space Age.

SEE ALSO: *Woman in the Moon* (1929), Origins of the Saturn V Moon Rocket (1930–44)

1870s

A **statue** of Russian aeronautical scientist Konstantin Eduardovich Tsiolkovsky, located in Moscow.

SCIENTISTS CONSIDER
LUNAR ORIGINS

ANAXAGORAS ONCE PROPOSED THAT THE Moon had begun as a rock flung off from the Earth, but real scientific consideration of lunar origins began in 1873. In that year, Édouard Albert Roche (1820–83) proposed *coaccretion*, the hypothesis that the Moon and Earth coalesced together from the same materials. Roche's hypothesis predicted that the Moon should be chemically identical to the Earth, but NASA's Apollo missions would reveal the Moon as deficient, compared with Earth, in iron and volatile materials. Roche also fell short of explaining the inclination of the Moon's orbit relative to the Earth's; but just prior to the Space Age, his idea enjoyed some notable support from the American planetary scientist Ralph Belknap Baldwin (1912–2010).

Back in 1739, Edmund Halley had confirmed the secular acceleration of the Moon, related to tides and Earth's rotation, which meant that the Moon should be getting gradually farther away from the Earth. In the 1890s, British-American astronomer Ernest William Brown (1866–1938) would provide additional details connecting lunar motion to Earth's spin; but while Roche was doing his work in France, the English geophysicist-astronomer George Darwin (1845–1912)—son of Charles—was confirming that the Moon was getting more distant. In 1878, this led Darwin to posit with Osmond Fisher (1817–1914) that the Moon actually had started on Earth. Some 56 million years ago, by their calculations, the Moon had split off from Earth, leaving the Pacific basin as a scar. Known as the fission hypothesis, the Darwin-Fisher idea required Earth to have been spinning much faster in the past than planetary scientists think it actually did.

Then, in 1909, the American astronomer Thomas Jefferson Jackson See (1866–1962) envisioned the Moon forming far away, then being captured by Earth's gravitational field. Nobel Prize-winning physical chemist Harold Urey (1893–1981) later would gravitate to Jackson's capture hypothesis. However, analysis of lunar rocks collected on Apollo missions would show that the Moon and Earth matched in their ratios of oxygen isotopes, meaning that chemically the two worlds are not so different after all. The mystery of the Moon's beginnings would continue.

SEE ALSO: Formation of the Moon (4.5 Billion Years Ago), Elucidating Lunar History (1970s–80s), Preparing for New Missions (2018)

NASA lab technicians study a lunar rock, brought to Earth from the Fra Mauro region of the Moon by the crew of Apollo 14.

THE MOON PROVES
GENERAL RELATIVITY

ALBERT EINSTEIN (1879–1955) DEVELoped his general relativity theory from 1907 to 1915, but he published milestones on the topic beginning in 1911. This boosted his academic recognition; but his theory, which redefined gravity as a bending force replacing the Newtonian pulling force, ignited a backlash. Several scientists opposed Einstein, one being Thomas Jefferson Jackson See, the astronomer who had proposed lunar capture to explain the Moon's origins.

Einstein realized that astronomers could actually assist him, because general relativity predicted that the Sun's gravity should bend light traveling from another star. Astronomers could observe stars appearing near the Sun's position in the sky and check for deflection of the star position compared to where the same star appeared at other times in the year. Such deflected starlight would be visible only during a full solar eclipse.

Einstein arranged for a young German astronomer, Erwin Finlay-Freundlich (1885–1964), to travel into Crimea, in Russian territory, where the eclipse would be visible on August 21, 1914. Collaborating with Freundlich, the American astronomer William Wallace Campbell (1862–1938) headed to the outskirts of Kiev with very special equipment. While the astronomers were traveling, World War I started on July 28, and Germany and Russia were enemies on August 1. Being a German with telescopic cameras, Freundlich was arrested for espionage. Meanwhile, clouds ruined Campbell's observations, and his equipment was confiscated. Successful observation of the eclipse would have discredited Einstein, however, as he discovered an error in his calculations. This led him to rework general relativity, but he still needed evidence from a solar eclipse.

In 1918, Campbell's measurements during another eclipse showed a lack of deflection, but he had used inferior instrumentation, as Russia still held his advanced equipment. He nearly published his findings in 1919, when English astronomer Arthur Eddington (1882–1944) was monitoring another eclipse, with good equipment, from the jungles of Africa. Eddington found that solar gravity did indeed deflect starlight precisely as general relativity predicted. In 1922, Campbell and others utilized still another solar eclipse to corroborate Eddington's findings.

SEE ALSO: The Moon Inspires Isaac Newton (Late 17th Century), Scientists Consider Lunar Origins (1873–1909)

Albert Einstein (top left) and Arthur Eddington (bottom left) pose with colleagues, 1923.

FIRST LIQUID-FUELED ROCKET

N THE FALL OF 1899, ROBERT GODDARD (1882–1945) climbed a cherry tree in Worcester, Masschusetts. Looking across a field, he imagined a craft ascending on a plume of hot gases, setting out for Mars. The seventeen-year-old Goddard knew about one type of rocket: fireworks, essentially tubes filled with solid fuel. He enjoyed launching fireworks each Fourth of July, but *The War of the Worlds*, a novel by H. G. Wells, had left Goddard pondering spaceflight. Now, the vision in the tree had him thinking that small rockets might be scaled up.

By 1914, Goddard was a physics professor and patenting an idea for liquid propellants. This was a conceptual challenge: liquid fuels were known to ignite all at once, while rockets needed their fuel to burn gradually. Goddard's solution was liquid oxygen, used both as an oxidizer delivered gradually to the engine, and as a coolant to help keep the engine from exploding.

A small grant from the Smithsonian Institution in 1917 sent Goddard on course to launch the first liquid-fueled rocket on March 16, 1926. Launched from Auburn, Massachusetts, the device boosted Goddard's public reputation, which had suffered since 1920, when in a Smithsonian article, "A Method for Reaching Extreme Altitudes," he had suggested that rockets could reach the Moon. A *New York Times* editorial later that year had ridiculed Goddard on the grounds that rockets must push against air, and thus would not work in space. The editorial had misconstrued Newton's third law of motion and was but one of numerous media insults.

But the 1926 launch led to more rockets and more launches that got the attention of aviator Charles Lindbergh (1902–74). This led to a grant from the Guggenheim family that enabled Goddard to send a rocket to an altitude of 2.7 kilometers by 1937, and ultimately to solve all basic issues of rocket flight. Along the way, he also anticipated ion propulsion (electric thrusting) for deep space missions. The *Times* would apologize for its editorial, only in 1969—24 years after Goddard's death, and one day after rocket engines, thrusting in space, had launched Apollo 11 toward the Moon.

SEE ALSO: *Woman in the Moon* (1929), BIS Lunar Spaceship Design (1938), Origins of the Saturn V Moon Rocket (1930–44)

Dr. Robert H. Goddard stands next to an early model rocket fueled by liquid propellant, which he launched on March 16, 1926, the day this photo was taken. He was the first to launch a rocket of this kind.

WOMAN IN THE MOON

JULES VERNE'S *DE LA TERRE À LA LUNE* (*From the Earth to the Moon*) inspired not only Konstantin Tsiolkovsky, but also two other visionaries who, together with Tsiolkovsky and Robert Goddard, laid out the principles of astronautics. One such visionary was Verne's compatriot Robert Esnault-Pelterie (1881–1957). Esnault-Pelterie, like Tsiolkovsky, recognized that a cannon launch would crush lunar explorers. Thus, by 1931, he was experimenting with liquid-fueled rockets in France. Eventually, he proposed that craft in deep space could run on nuclear power. Despite his vision, Esnault-Pelterie was unable to spark government interest in astronautics, just as Goddard was unable to do the same in America.

The other Verne-inspired visionary was the Romanian-born, German-educated physicist Hermann Oberth (1894–1989). Having entered medical study at age sixteen, Oberth expected to work as a physician, but he was drafted into World War I. While assigned to a medical unit at a hospital in his native Transylvania, he conducted experiments on simulated weightlessness, and then resumed rocketry experiments that he had begun in childhood.

Oberth was testing a liquid-fueled rocket engine in 1929, by which time he was committed to a strategy, one that Tsiolkovsky and Goddard also emphasized would be key to practical spaceflight: multi-stage rockets. On top of his research, Oberth actively popularized the vision of space exploration, including the staging idea. This led the German film industry to seek Oberth as a scientific adviser for a 1929 silent film, *Frau im Mond* (*Woman in the Moon*), about a group of scientists and businessmen who travel to the moon seeking gold. Oberth inserted in the film the multi-staging idea, along with several other features that would become familiar to human spaceflight. These features included a vehicle-assembly building, from which the rocket craft was transferred to a launch pad; the use of water beneath the engines to dampen launch vibrations; horizontal positioning of astronauts to increase tolerance to G-forces during launch; and the countdown. Considering certain details of the fictional rocket too similar to features of their real but secret rocket program, the Nazis banned the film in Germany from 1933 to 1945.

SEE ALSO: Verne Inspires the Father of Astronautics (1870s), BIS Lunar Spaceship Design (1938), Origins of the Saturn V Moon Rocket (1930–44)

A still from the German silent film *Frau im Mond,* showing characters disembarking from their rocket.

BIS LUNAR SPACESHIP DESIGN

1938

ROBERT GODDARD COULDN'T CONVINCE the United States government of the potential of liquid-fueled rockets, yet received technical questions from engineers in Nazi Germany. The Germans may have considered Goddard a potential collaborator, given his association with Charles Lindbergh, who, by the late 1930s, was an outspoken white supremacist, anti-Semite, and Nazi sympathizer. Goddard avoided the Germans, but they studied his journal articles in detail. With Wehrmacht funding, they expanded on Goddard's innovations and added their own ideas to produce the Aggregat rockets, with the goal of carrying bombs.

With the US ignoring rockets and the Nazis developing rocket weapons, the prospects for space exploration looked bleak. Even worse, the future in general looked bleak, with war looming. And yet, in 1938, the British Interplanetary Society (BIS), founded just five years earlier, was paying attention both to Goddard's vision of the future and to the German technical innovations. Engineers in the BIS were thinking about multiple-stage rockets, which Goddard had described in technical detail, and which Hermann Oberth had gone so far as to popularize through his technical-advising on the film *Woman in the Moon.*

Using multiple stages with liquid-fueled engines might increase performance further, but the BIS design team had an additional idea: engine clusters. To achieve the lifting capability and velocity change needed to reach the Moon, the team designed five lower stages, each with 168 engines, and a sixth stage with 45 engines. The combined thrust possible from the engine clusters plus the staging made human spaceflight look realistic; but in place of Goddard's dream of sending humans to Mars, the BIS team created a conceptual design with a more modest goal: a spacecraft that could carry three people to the Moon and return them safely to Earth after a fourteen-day stay.

The BIS engineers recognized numerous technological obstacles to their design, particularly engine clustering, which introduced problems that would not be addressed until the 1950s. But the point of the project was to show that a Moon flight was feasible and economically viable. Little did they know that such flights lay just three decades into the future.

SEE ALSO: First Liquid-Fueled Rocket (1926), BIS Lunar Spaceship Design (1938), Origins of the Saturn V Moon Rocket (1930–44)

A painting of the BIS Lunar Spaceship by Ralph Andrew Smith, who served as president of the BIS in 1956–57.

ORIGINS OF THE SATURN V MOON ROCKET

Konstantin Tsiolkovsky, Robert Goddard, Robert Esnault-Pelterie, and Hermann Oberth established the principles of astronautics, but it was a second generation of space pioneers that made lunar missions a reality. Around 1930, teenage space enthusiast Wernher von Braun (1912–77) joined Oberth's research team. While pursuing a degree in aerospace engineering, von Braun was recruited into a military rocket research program. Directed by Walter Dornberger (1895–1980) within the Reichswehr, or armed forces of the Weimar Republic, the program funded von Braun's doctoral studies when the Nazis seized control of Germany in 1933 and the team was tasked to develop the Aggregat rockets.

Under the control of the *Wehrmacht*—Nazi Germany's military—from 1935, funding increased substantially, and later the researchers relocated to Peenemünde on the Baltic coast. A high-altitude rocket called the A-4 first flew in October 1942, leading Dornberger to declare that the era of space travel had begun. He was correct, as there would be a direct evolution from this rocket to the Saturn V booster that would send astronauts to the Moon. Much sooner, in June 1944, an A-4 would reach 108 miles (174 kilometers) in altitude, true flight through outer space.

The Wehrmacht was interested purely in bomb delivery. Seeing this as his only pathway to space, von Braun joined the Nazi party in 1938, and two years later he was "invited" to become an SS officer. In September 1944, bomb-carrying A-4s—renamed V-2s, "V" standing for vengeance—started hitting London. Still believing that war would give way to space exploration, von Braun remarked that "the rocket worked perfectly, but it landed on the wrong planet."

Other than concluding from von Braun's work and various testimonies that he did not admire militarism or Nazi ideology, historians have struggled to work out the state of his conscience during the Nazi period. Aerospace historian Michael Neufeld has described von Braun as having "sleepwalked into a Faustian deal," but what's clear is that there were other V-2 designers who embraced Nazism with enthusiasm. And some of them would help in jump-starting the American space program.

SEE ALSO: First Liquid-Fueled Rocket (1926), *Woman in the Moon* (1929), Operation Overcast (1945), A New Discovery and a New Agency (1958–59), Planning Lunar Missions (1962)

1930-44

Dr. Wernher von Braun stands next to the engines of the Saturn V.

OPERATION OVERCAST

N 1945, THE US BEGAN OPERATION OVERCAST to locate German technical specialists in various fields. Authorized by President Harry Truman (1884–1972) and later renamed Operation Paperclip, the project transported 1,600 technical experts to US soil, including Wernher von Braun. Knowing that the Allies and Soviets were approaching, von Braun and his colleagues walked a tight line between orchestrating a surrender to the Americans and responding to SS commanders, who had ordered them to destroy their rocket documents. Risking execution, von Braun and his assistant Dieter Huzel (1912–94) hid rocket blueprints. Meanwhile, President Truman ordered that no war criminals be included among the recruited German specialists.

Offsetting the legacy of Truman's order is an ugly reality that still haunts the history of lunar exploration. Hoping to out-compete the USSR, US Intelligence avoided scrutinizing the backgrounds of captured Germans with special skills, even as others were being prosecuted at the Nuremberg Trials. Back in 1943, the Allies had bombed the Peenemünde rocket facility, and V-2 production had moved southward, to Mittelwerk, a factory consisting of tunnels carved into a mountain.

Mittelwerk had put out rockets that killed some 9,000 people in London and Antwerp, but it is estimated that as many as 20,000 people—slaves from the Dora concentration camp—died in connection with V-2 rocket production at the factory under horrid conditions. The British captured Walter Dornberger and investigated his role in Mittelwerk for possible war crimes; they later released him to work for the US Air Force. Investigation in the 1980s would reveal that the idea to use slaves had come from V-2 engineer Arthur Rudolph (1906–96), who had joined the Nazi Party enthusiastically in 1931. But in 1945 Rudolph was whisked to the US with von Braun and others of the V-2 team. From 1963 to 1968, Rudolph would serve as project director for NASA's Saturn V program, but in 1984 he would relinquish his US citizenship and flee the US to avoid war- crimes prosecution. Meanwhile, von Braun had affixed his signature to the Mittelwerk slavery plan and later admitted to having witnessed the conditions in the tunnels.

SEE ALSO: Origins of the Saturn V Moon Rocket (1930–44), A New Discovery and a New Agency (1958–59), Planning Lunar Missions (1962)

This tunnel in Mittelwerk led to a storage area that housed components needed to produce V-2 rockets.

NEW UNDERSTANDING
OF CRATERS

N 1829, Franz von Gruithuisen proposed meteorite impacts as the cause of lunar craters. Later, scientists said that the Moon's craters must be volcanic calderas. Then, in 1891, Daniel Barringer (1860–1929) revived the impact idea around the time that he was studying what is now called Barringer Crater. Located in northern Arizona, it's also called Meteor Crater, because Barringer convinced other geologists that it had resulted from an impact. Meanwhile, Grove Karl Gilbert (1843–1918) of the United States Geological Survey (USGS) insisted that Barringer Crater was a caldera. Geologists were fully divided on this matter by the 1940s, but World War II shook things up.

Like the push to advance rocket technology, the effort to understand cratering developed in connection with humanity's new weapon, the atomic bomb. Scientists were put to work understanding the biological and geophysical effects of nuclear blasts. It was in this setting that Eugene Merle Shoemaker (1928–97) began his graduate studies, with USGS sponsorship to study nuclear blast craters at the Nevada nuclear test site. Just 19 years of age upon receiving his bachelor's degree from the California Institute of Technology (Caltech)

in 1948 and beginning a Master of Science program that he would complete at the same institution a year later, Shoemaker earned a PhD from Princeton University over the course of a decade while working as a USGS scientist.

At the Nevada test site, Shoemaker found ejected material containing shocked quartz forming a ring around the blast craters. Microscopic analysis showed the quartz to be a particular structure resulting from intense pressure, and Shoemaker found similar features in quartz from Barringer Crater. Volcanism could not produce such effects, but a meteorite impact could. Shoemaker thus concluded that Barringer Crater had formed from an iron-rich meteorite. He reasoned that impacts must similarly have carved out craters on the Moon. This idea partly overlapped with a hypothesis put forward by planetary scientist Ralph Belknap Baldwin, who was also making a case that lunar craters resulted from impact events, though through a slightly different mechanism. It also put Shoemaker on the ground floor of a new science called astrogeology.

SEE ALSO: Another Doctor Turns His Eyes to the Moon (1824), Astrogeology (1964–65), Lunar Prospector and Surface Ice (1998)

The discovery that the Barringer Crater in northern Arizona resulted from a meteorite impact, rather than volcanism, helped scientists realize that lunar craters were formed from similar impacts. Also called the Meteor Crater, it measures about three-quarters of a mile wide (1.2 km) and 560 feet (170 m) in depth.

SPUTNIK

THE UNITED STATES AND THE SOVIET Union both initiated space programs as a result of rocket engineers developing missiles for military needs, while urging their governments to allow them to place a satellite into orbit—a prerequisite for any other space-flight achievement, including flights to the Moon. Along with engineers from war-torn Germany, the USSR had Sergey Korolyov. In 1933, Korolyov had launched the USSR's first liquid-fueled rocket, but he'd been imprisoned in a gulag in 1938. The Soviet government had released him near the end of World War II, so he could work with German engineers after the war to build a Soviet rocket program.

By 1953, the US and USSR both had hydrogen bombs, and the Soviets wanted a missile that could carry a five-ton warhead across continents. Hoping that meeting the military's needs would set him up to launch a satellite, Korolyov proposed a novel rocket, the R–7, propelled by dozens of clustered engines, somewhat reminiscent of the engine-clustering that the British Interplanetary Society had envisioned for its fanciful 1938 moon ship.

The R–7 worked, and Korolyov's next challenge was to sell the satellite idea to Soviet leader Nikita Khrushchev (1894–1971) based on the possibility that such a craft could be used for spying. Unaware that the US was already developing spy satellites (despite not yet having the capacity to launch them), Khrushchev was initially skeptical. But he gave Korolyov the green light to launch a satellite for the publicity value. Sputnik, the basketball-sized satellite that Korolyov launched on October 4, 1957, contained only a transmitter to prove that it was in orbit. But Americans woke up both fascinated and concerned. Some journalists speculated that the orbiter might be a nuclear bomb. Then, a month later, Korolyov launched Sputnik 2, carrying a dog, Laika, who survived five hours in orbit before succumbing to heat exhaustion. Laika died a celebrity, but the world knew not of Korolyov. For his protection, his identity was a tightly guarded state secret.

SEE ALSO: Verne Inspires the Father of Astronautics (1870s), BIS Lunar Spaceship Design (1938)

Laika, a stray who had been living on the streets of Moscow, became the first animal to orbit Earth.

EXPLORER I

I N 1950, THE US ARMY RELOCATED WERNHER von Braun's team of German rocket designers from Texas to the Redstone Arsenal in Huntsville, Alabama. Expanded to include some American engineers with Major General John Medaris (1902–90) as commander and von Braun as technical director, the group became the Army Ballistic Missile Agency (ABMA) in 1956. The task was to upgrade the V-2 rocket into a family of ballistic missiles, called *Redstone*, for military use. However, in 1954, the Army proposed utilizing Redstone to launch satellites. This was to be part of an Army–Navy program called Project Orbiter, but the Navy also had its own satellite-launching project, called *Vanguard*.

To get public support for space exploration, von Braun collaborated with Walt Disney Studios to make animated videos of his vision of spaceflight of the future. With the demonstration of a Redstone-class rocket called Jupiter-C, the ABMA declared its readiness to launch a satellite in mid-1956, but this was a collaborative effort. Under the direction of William Pickering (1910–2004), the Jet Propulsion Laboratory (JPL) in California was building the satellite, plus three upper stages needed to boost it into orbit, and this took

time. Meanwhile, deciding that an American rocket, not a German V-2 derivative, should initiate the Space Age, the Eisenhower administration allowed the Navy to make the first launch attempt. The Vanguard launch was scheduled to occur from Cape Canaveral, Florida in December 1957. That was a month after Sputnik 2 carried the dog Laika to space, and the satellite atop the Vanguard was tiny. Nikita Khrushchev called it a "grapefruit," but a greater insult was to come from the rocket itself. On live television, the Vanguard collapsed and exploded after rising just a meter.

On January 31, 1958, the ABMA-JPL team launched the satellite Explorer I into an orbit higher than either of the two Soviet missions. Featuring a cosmic ray detector designed by University of Iowa physicist James Van Allen (1914–2006), Explorer I detected radiation particles trapped in certain regions and altitudes by Earth's magnetic field. This led Van Allen to propose the existence of radiation belts that later would be vital to the planning of piloted lunar missions.

SEE ALSO: A New Discovery and a New Agency (1958–59), Apollo Biostack (1972)

The Explorer I, launched on January 31, 1959 from Cape Canaveral, Florida. It was the first American satellite to orbit Earth.

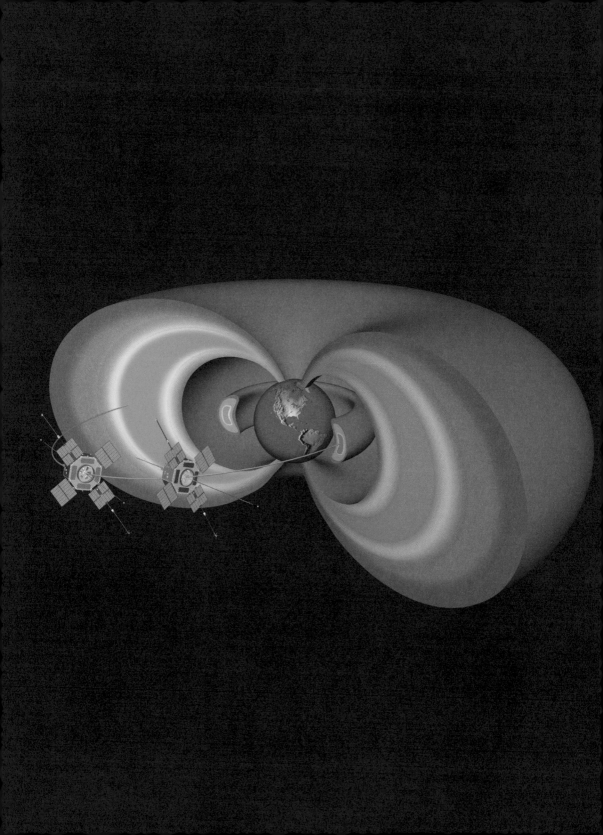

A NEW DISCOVERY
AND A NEW AGENCY

THE INTERNATIONAL GEOPHYSICAL YEAR (July 1957 through December 1958) included two developments that would prove vital to human lunar missions. One was the discovery and early study of the Van Allen radiation belts. The other was the formation of the National Aeronautics and Space Administration (NASA).

Resulting from the Earth's magnetosphere deflecting deep-space–charged particles away from Earth, but then trapping them at high altitudes, the belts are named for James Van Allen, who discovered them with the cosmic-ray sensors of Explorer 1, Explorer 3, and Explorer 4. Launched during 1958, these probes traversed what is called the inner Van Allen Belt, a persistently dumbbell-shaped region of trapped space-radiation particles. An outer Van Allen Belt also is present constantly, but with slight shape variation. (In 2013, NASA's two Van Allen Probes would reveal a third belt that appears only when the Sun produces what are called solar particle events.)

The belts meant that human lunar flights would have to avoid regions where charged particles were most heavily concentrated. Explorers 1, 3, and 4 did not reach the outer belt, but the Soviet Sputnik 3 detected it, and, unknown at the time, Sputnik 2 had flown through it. In December 1958, attempting to reach the Moon, the American Pioneer 3 also confirmed the presence of the outer belt, as did the Soviet Luna 1 on January 4, 1959. This craft flew past the Moon and then became the first human-made craft to orbit the Sun.

Also important to lunar exploration was the formation of NASA. In the early days of the Space Race, the US Army, Navy, and Air Force launched space missions separately. But this changed when NASA was formed on October 1, 1958. Along with absorbing the National Advisory Committee for Aeronautics (NACA), the new agency began managing the Caltech-contracted Jet Propulsion Laboratory (JPL), and absorbed Navy and Army astronautics centers. This included the Army Ballistic Missile Agency (ABMA), soon to be designated the Marshall Spaceflight Center. Now working as a NASA employee, Wernher von Braun began developing a new family of rockets—the Saturn rockets—that would dwarf his Redstone-Jupiter series and ultimately launch human missions to the Moon.

SEE ALSO: Improving Instruments Advance Lunar Astronomy (18th Century), Apollo Biostack (1972)

This artist's conception shows both the inner and outer Van Allen radiation belts. The outer belt varies in shape depending on solar activity. It is now known that there is also a third belt, appearing only during solar particle events.

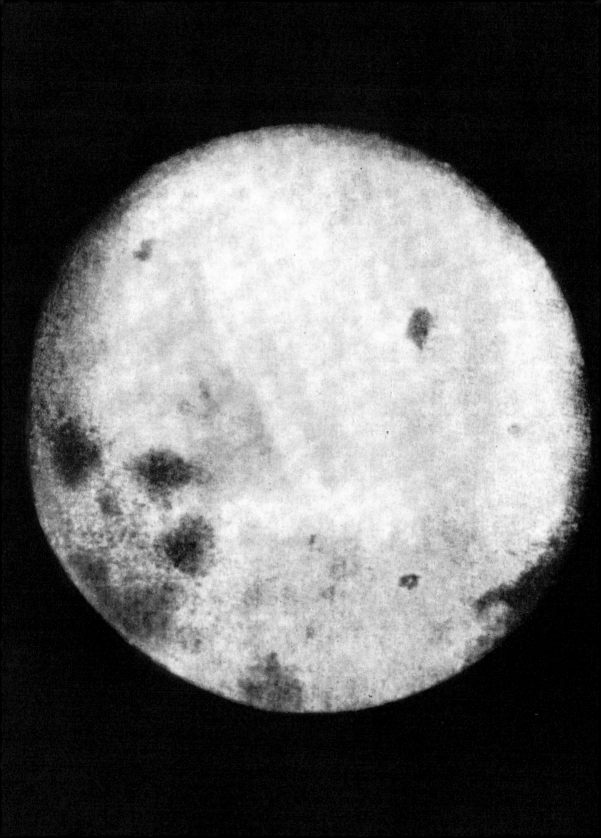

FIRST PICTURES OF THE MOON'S FARSIDE

FTER THE USSR AND US HAD PLACED satellites into orbit, there was talk about the superpowers being in a kind of Space Race, with the Moon as the expected finish line. When a Soviet probe called Luna 2 impacted the Moon in September 1959, Soviet leader Nikita Khrushchev was in the US, visiting an Iowa sausage-packing plant, where he made sure to milk the achievement for all possible publicity. "We have beaten you to the Moon," he declared, "but you have beaten us in sausage-making."

Only three weeks later, the Soviet Union enjoyed an even greater scientific victory. On October 6, 1959, the probe Luna 3 flew around the Moon, beaming back photographs from the lunar farside, a region that had been invisible to humans for the entire course of history. The quality of the Luna 3 photographs was limited, but one thing was striking: the far side looked deficient in maria, the dark "seas" that are so characteristic of the side that we see from Earth.

Luna 3 was a scientific milestone, but the USSR and the US both had programs underway to get humans into orbit. Powerful enough to deliver an H-bomb across continents, Sergey Korolyov's R–7 needed only some modifications to carry a human into orbit. Meanwhile, the US was developing the Atlas rocket for dual use—bomb delivery, and human orbital flight. By 1960, Korolyov's team had built Vostok, a capsular craft designed to carry a human. However, the R–7 rocket depended on military funding, and engine designer Valentin Glushko (1908–89) was offering the military a new engine using hypergolic fuel and oxidizer, chemicals that ignite spontaneously when combined. With safety precautions, hypergolics would later enable NASA's Gemini and Apollo missions, but they sounded dangerous for human flight in 1960, as they produced fumes destructive to the lungs. However, since Glushko's design would enable rapid emergency launches, it would render the R-7 obsolete for weapons delivery. Facing a potential funding crisis, Korolyov accelerated his efforts to demonstrate that a Vostok, lofted by an R–7, could carry large mammals into orbit and return them safely to Earth.

SEE ALSO: Sputnik (1957), Beginnings of Lunar Science (1964)

A photograph of the Moon's farside from Soviet probe Luna 3. This was the first time that any human had seen the farside.

HUMANS ENTER SPACE

1961

FACING POTENTIAL BUDGET CUTS, SOVIET spacecraft designer Sergey Korolyov was eager to show that his new Vostok capsule was safe to carry a human into space atop an R–7 rocket. In 1957, the dog Laika had flown in Sputnik 2 with no chance of survival, and had suffered a grueling death from heat exhaustion only hours into the flight. With survival as the goal in 1960, Korolyov began launching canine cosmonauts on orbital flights, some in pairs, some alone. These dog flights had a roughly 50 percent survival rate.

Meanwhile, Korolyov was overseeing the cosmonaut selection program, which was so secret that the candidates did not even know what they were competing for until the winners were informed. On the other side of the globe, however, NASA announced its first human space program, Project Mercury, in a press conference with America's first seven astronauts in full public spotlight. Orbital flight was the objective, but the Atlas rocket that was supposed to propel Mercury had suffered setbacks, and intelligence reports said that the Soviets were getting close to a human launch.

Consequently, NASA opted to fit the first Mercury capsules to a Redstone rocket. With hundreds of modifications, Wernher von Braun was able to make the Redstone human-safe, though it could launch a Mercury capsule only on a suborbital spaceflight. While launching a chimpanzee, however, a rocket malfunction caused excessive acceleration that almost pushed the astrochimp into G-LOC—gravity loss of consciousness, a life-threatening situation. The astrochimp survived this and other rough phases of the flight, and potentially so could human astronauts. Moreover, von Braun and his team were able to diagnose and solve the problem. Consequently, flight director Christopher Kraft (b. 1924)—a pioneer in spaceflight operations and the inventor of the concept of Mission Control—was convinced that it was safe to attempt a human launch. Von Braun's team wanted another test flight, however, and this delayed the first human Mercury flight beyond April 12, 1961. On that day, cosmonaut Yuri Alexeyevich Gagarin (1934–68) became the first human to fly in space, making a full orbit on a flight lasting 108 minutes. With an apogee (highest point) of just 177 nautical miles, Gagarin's orbit had taken him just slightly above Earth's atmosphere.

SEE ALSO: Sputnik (1957), Explorer I (1958), An American in Space (1961)

Soviet cosmonaut Yuri Alexeyevich Gagarin wearing his spacesuit, 1961.

AN AMERICAN IN SPACE

"I believe that this nation should commit itself to achieving the goal, before this decade is out, of landing a man on the Moon and returning him safely to the Earth."

—John F. Kennedy, "Special Message to the Congress on Urgent National Needs," May 25, 1961

O N MAY 5, 1961, NASA LAUNCHED ASTRO- naut Alan Shepard (1923–98) on a suborbital flight lasting 15 minutes and 28 seconds. Shepard's mission featured 6 minutes of weightlessness and a view of the Earth from space that only cosmonaut Yuri Gagarin had seen.

Gagarin was now a celebrity, but the same was not true of Sergey Korolyov, chief of the Special Design Bureau 1 (OKB–1) and master- mind of the Soviet space program. Enforced by the Committee for State Security (KGB), Korolyov's identity was a state secret.

Also unknown to the world were the power struggles within the Soviet space program. Personal tensions had always been strong between Korolyov and engine specialist Valentin Glushko, but these had exacerbated by 1960 because of disagreements over engine strategies. Complicating the situation further,

Korolyov wanted desperately to launch a human mission to the Moon; but another engineer, Vladimir Chelomey (1914–84), was given control of lunar exploration because he had earned the favor of Soviet leader Nikita Khrushchev.

Meanwhile, a military rocket had been set up for testing on October 24, 1960, loaded with Glushko's hypergolic fuel. In the midst of rushed, improvised repairs, the second stage (while still on the ground) had ignited, killing the military missile program commander and some eighty or ninety (possibly more) senior rocket engineers. Keeping the disaster a secret, the USSR preserved the image that it was years ahead of the US in space. But at NASA's Marshall Spaceflight Center in Alabama, Wernher von Braun was almost ready to test Saturn I, a multi-stage rocket with more lift- ing power than Korolyov's R-7. Knowing this, US President John F. Kennedy (1917–63) soon delivered a declaration to a joint session of Congress, committing the US to reaching the Moon by the end of the "decade" (technically December 31, 1970).

SEE ALSO: Sputnik (1957), Explorer I (1958), The Moon Speech at Rice Stadium (1962)

These seven astronauts were selected for Project Mercury, whose aim was to put a person into orbit and return him safely. Back row, from left to right: Alan Shepard, Virgil "Gus" Grissom, and L. Gordon Cooper, Jr. Front row, from left to right: Wally Schirra, Donald "Deke" Slayton, John Glenn, and Scott Carpenter.

SATURN-NOVA

COMPARISON

Spacecraft

18'-4" dia.

270'

33' dia.

21'-5" dia.

Spacecraft

18'-4"dia.

125'

Spacecraft

22' dia.

40' dia.

280'

50' dia.

C-1

C-5

NOVA

M-MS-G-36-62, Apr. 11

PLANNING LUNAR MISSIONS

KENNEDY'S TIMEFRAME WAS SO TIGHT that some space experts, notably Christopher Kraft, the inventor of Mission Control, expressed concern about the president's state of mind. Meanwhile, in a test in October 1962, the first stage of Wernher von Braun's Saturn I rocket delivered more thrust than any previous booster. This was a stepping stone toward the Saturn V, but von Braun had an even more massive rocket on the drawing board, called Nova.

In a strategy called direct ascent, Nova would deliver a 40- to 50-ton spacecraft to the lunar surface, with enough fuel for a return flight. Along with concerns that the ship would be so tall that astronauts might have trouble knowing their altitude just before touchdown, there were other drawbacks. Each Nova stage had to be built in a new factory, and the Nova would not be ready until well into the 1970s. In comparison, Saturn V required a new factory only for its first stage, known as the S-1C, and might be ready in only a few years.

Since the Saturn could not support a direct-ascent round trip, early mission planners actually discussed an idea of NASA sending an astronaut on a one-way lunar mission before the Soviets had a chance to get there first. Supply ships would then maintain the astronaut, until a better rocket could be sent to bring him home. Another option, Earth orbit rendezvous (EOR), would require, instead of Nova, two Saturn Vs to lift a lunar craft and a refueling craft that would rendezvous in low Earth orbit (LEO).

Von Braun came to favor EOR. Risking his career by going several levels over his own head and writing directly to NASA associate administrator Robert Seamans (1918–2008), aerospace engineer John Houbolt (1919–2014) championed a different architecture. Known as lunar-orbit rendezvous (LOR), Houbolt's plan involved a mother ship and a lightweight lunar craft traveling together from LEO to lunar orbit, from which only the lightweight craft would descend. This mission could be launched on a single Saturn V. In 1962, NASA Administrator James Webb (1906–92) announced that LOR would be the strategy for lunar missions. The ability to rendezvous and dock spacecraft was now critical to the program.

SEE ALSO: Origins of the Saturn V Moon Rocket (1930–44), Saturn Architecture Takes Shape (1963–64)

A 1962 concept drawing of a Nova rocket (right) that compares the sizes of its components to those of the Saturn C–1 (left) and Saturn C–5 (center).

SEAL OF THE PRESIDENT OF THE UNITED STATES · E PLURIBUS UNUM ·

THE MOON SPEECH
AT RICE STADIUM

1962

N AUGUST 1962, THE USSR HIT NEW SPACE milestones in orbiting two Vostok craft simultaneously, and showing that humans could tolerate almost four days of weightlessness. But with the Atlas rocket now operational, NASA had two human orbital missions under its belt, plus a swelling budget to support President John F. Kennedy's mandate for piloted lunar missions. In September, at Rice University, JFK provided historical perspective for the project ahead:

> No man can fully grasp how far and how fast we have come, but condense, if you will, the 50,000 years of man's recorded history in a time span of but a half-century. Stated in these terms, we know very little about the first 40 years, except at the end of them advanced man had learned to use the skins of animals to cover them. . . . Only five years ago man learned to write and use a cart with wheels. . . . Last month electric lights and telephones and automobiles and airplanes became available. Only last week did we develop penicillin and television and nuclear power, and now if America's new spacecraft succeeds in reaching Venus, we will have literally reached the stars before midnight tonight. . . .

> We set sail on this new sea because there is new knowledge to be gained, and new rights to be won, and they must be won and used for the progress of all people. For space science, like nuclear science and all technology, has no conscience of its own. Whether it will become a force for good or ill depends on man. . . .

> We choose to go to the Moon. We choose to go to the Moon in this decade and do the other things, not because they are easy, but because they are hard, because that goal will serve to organize and measure the best of our energies and skills, because that challenge is one that we are willing to accept, one we are unwilling to postpone, and one which we intend to win, and the others, too.

—Excerpted from a speech by John F. Kennedy, delivered September 12, 1962, at Rice University in Houston, Texas

SEE ALSO: Sputnik (1957), An American in Space (1961), Saturn Architecture Takes Shape (1963–64)

President John F. Kennedy addresses a crowd at Rice University's football stadium to outline the importance of reaching the Moon and build support for the United States' space program.

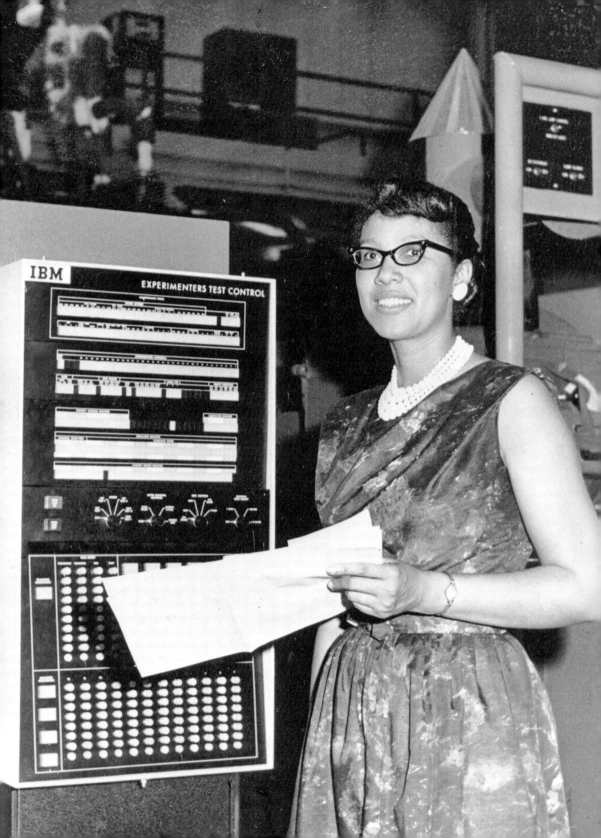

HUMAN COMPUTERS

N 1963, THE USSR HAD ONE FEMALE CIVILIAN cosmonaut orbiting Earth and subsequently marching in parades and appearing in *Life* magazine. In 1963, NASA's astronauts were all white, male, military test pilots, and yet the agency had dozens of women working behind the scenes, many of them African American.

One of 400 candidates applying to become the first woman in space, Valentina Tereshkova (b. 1937) had topped out to fly Vostok-6 over other women, because she was an excellent parachute jumper and because her work for the Communist Party gave her the right ideological stuff. Working as human "computers" since the days of NASA's predecessor, the National Advisory Committee for Aeronautics, the women of Project Mercury were chosen because they could do calculus, three-dimensional analytic geometry, and other mathematics essential for the Atlas rocket to send each Mercury spacecraft into orbit, and for retrorockets to fire in the correct direction, for a precise amount of time, to bring the astronaut home. Too long, and the spacecraft would burn up reentering the atmosphere. Too short, and the spacecraft would bounce off the upper atmosphere, never to return.

The human computers got no publicity or glory, but the astronauts knew that their lives depended on these women, even though NASA had begun calculating flight parameters using a new technology—the electronic computer, a titan of vacuum tubes and spinning tape wheels filling two full rooms.

The transition to electronic computing had begun during Mercury-Atlas 6, the mission in which astronaut John Glenn (1921–2016) had become the first American to orbit Earth. "Better have Katherine verify it," Glenn had requested just prior to his February 1962 launch, referring to launch calculations and the computer Katherine Johnson (b. 1918).

By the end of 1963, Project Mercury was over and Americans were coping with the tragic loss of President Kennedy. But the lunar program was alive, and Johnson and her colleagues were transitioning from being computers themselves to programming the new computers that were to control the trajectories of a lunar-bound craft called Apollo.

SEE ALSO: Two's Company, Three's a Crowd (1964), Improving Space Capabilities (1965)

Melba Roy Mouton (1929-90) stands next to an electronic computer, 1964. Mouton led the group of female mathematicians working at NASA, also known as the computers. Prior to the arrival of electronic computing, the human computers conducted their calculations by hand.

SATURN ARCHITECTURE
TAKES SHAPE

RESIDENT KENNEDY NEVER WITNESSED a launch of the iconic Saturn V, which carried the Apollo lunar missions, but he did watch a test of its predecessor, Saturn I. This happened when he visited NASA's Marshall Spaceflight Center in Huntsville, Alabama, after setting the Moon landing goal before Congress in 1961. During his visit, JFK experienced instant rapport with Wernher von Braun. The rocket designer reassured the president that he had not gone off the deep end in committing to a piloted Moon landing.

The decision to now carry out a lunar-rendezvous strategy meant that von Braun could focus on the Saturn, and back-burner the even bigger Nova concept. Although more powerful than the Soviet R–7 rocket, the Saturn I first stage was basically a cluster of eight Redstone-Jupiter rockets and tanks. Von Braun's group had developed the Jupiter directly from the German V–2, meaning that the Saturn I first stage also was a V–2 derivative. Atop the first stage was an upper stage, called the S–IV, which by the mid-1960s had been modified substantially by replacing its six small engines with one big one, called a J2.

This transformed the S–IV into the S–IVB stage, which, mounted atop a modified Saturn I lower stage, produced the Saturn 1B.

The Saturn 1B could launch either an Apollo Command/Service module (CSM) or an Apollo lunar excursion module (LEM, the craft designed to ferry astronauts between lunar orbit and the lunar surface) into low Earth orbit (LEO). Early LEO test flights could utilize an early version of the CSM, called a block-I CSM. Lunar flights would require a more advanced block-II CSM and the three-staged Saturn V rocket. Known as the S–1C, the first Saturn V stage would have a cluster of five enormous F1 engines that together would consume fifteen tons of fuel and liquid oxygen per second. The middle stage, the S–II, would have five J2 engines. The third stage would be an S–IVB, like the Saturn 1B second stage. However, in the Saturn V version of this stage, the single J2 engine could be throttled down, and restarted later, in outer space.

SEE ALSO: Origins of the Saturn V Moon Rocket (1930–44), Planning Lunar Missions (1962)

President John F. Kennedy (right) receives an overview of the Saturn-IV launch system from Dr. Wernher von Braun (left) during a visit to Cape Canaveral on November 11, 1963, eleven days before Kennedy was assassinated.

TWO'S COMPANY, THREE'S A CROWD

FUNDED GENEROUSLY TO SEND HUMANS to the Moon, NASA had rocket designer Wernher von Braun developing the Saturn V, a launch vehicle that would make the Atlas rocket look like a toy. Meanwhile, the USSR was developing a new class of probes called *Zond* and a new, piloted craft called *Soyuz*, both envisioned for deep-space missions. This was possible because chief space designer Sergey Korolyov was skilled at selling ideas to the Soviet leadership, which was intrigued by the prospect of launching cosmonauts to orbit the Moon in time for October 1967, the 50th anniversary of the Bolshevik Revolution.

Nikita Khrushchev was not offering much funding, and Korolyov was struggling with other space experts over engine strategies for the needed launch vehicle. Even worse, Khrushchev was asking for unwise, spectacular missions, an example being Voskhod 1, which developed as a reaction to NASA testing its new Gemini spacecraft. Designed to let astronauts practice course corrections, spacecraft rendezvous, and docking, Gemini had the ability to change orbits and altitude, and an on-board computer. It also was expected to function in space for up to two weeks, making it a prelude to piloted lunar missions. One feature of the new program, however, was particularly obvious: the Gemini capsule had two seats.

Knowing that NASA was preparing to begin launching pairs of astronauts in Gemini beginning in early 1965, Khrushchev wanted the new Voskhod capsule launched by the end of 1964, plus he had an additional idea. To pull far ahead of NASA, he ordered Korolyov to send up three cosmonauts in the Voskhod 1 craft, which launched October 12, 1964. To make room for a third cosmonaut, Korolyov's team removed all safety equipment, including ejection seats and personal parachutes, the launch escape system, and spacesuits (which served as a backup for the possibility of cabin-pressure loss). The three cosmonauts were lucky to survive, but the same could not be said of Khrushchev's political career. Coincidentally, one day after the return of Voskhod 1, Khrushchev was ousted and replaced by Leonid Brezhnev (1906–82) as the most powerful man in the USSR.

SEE ALSO: Humans Enter Space (1961), Improving Space Capabilities (1965)

The three crew members aboard the Voskhod 1 return from their mission safely on October 13, 1964. From left to right: Vladimir Komarov (1927–67), Boris Yegorov (1937–94), and Konstatin Feoktistov (1926–2009).

BEGINNINGS OF LUNAR SCIENCE

OFFICIALLY, NASA'S FIRST SUCCESSFUL lunar probe was Pioneer 4, which launched March 3, 1959, and flew by the Moon, but this was a very distant flyby. At closest, the craft passed within 31,849 nautical miles (58,983 kilometers, or 36,650 statute miles) of the Moon. In terms of science, the mission took important measurements connected with the outer Van Allen Belt, a region of trapped radiation particles, before continuing into its own orbit around the Sun. However, the Soviet Luna 3 probe really won the show of that period, when it photographed the far side of the Moon seven months after Pioneer 4 had departed the Earth-Moon system. During the early 1960s, both the US and USSR made several attempts to deliver probes close to the Moon, or to the lunar surface, but there were several failures during launch, in low Earth orbit (LEO), or deeper in space.

From the late 1950s to the mid 1970s, in most cases, the name *Luna*, followed by a number, signified a Soviet probe that either made it to the Moon or did something useful related to the Moon. If a probe launched and got stuck in LEO, often it would be called *Cosmos*—followed by a number. And, if a Soviet rocket blew up attempting to launch a probe, there would be no name, because nobody would hear about the mission. NASA was thus reputed to suffer more lunar probe failures than the Soviet space program, particularly during the early years of a program called Ranger, conceived to obtain close-up images of the lunar surface to make it possible and safe for astronauts to land there later. From 1961 to 1962, the first five Rangers either did not make it to the Moon or lost contact with controllers. In January 1964, Ranger 6 was able to impact the Moon, but its cameras were destroyed in the process.

NASA's luck changed with Ranger 7, which impacted the Moon on July 31, 1964, and sent back thousands of television photographs. More Ranger missions and other unpiloted probes were gearing up for lunar studies later in the decade, but NASA was now also considering how astronauts might themselves engage in science on the lunar surface.

SEE ALSO: New Understanding of Craters (1948–60), A New Discovery and a New Agency (1958–59), Astrogeology (1964–65), Lunar Prospector and Surface Ice (1998)

The Ranger 7 transmitted more than 4,300 images before crashing into the Moon's surface.

ASTROGEOLOGY

HAVING DEMONSTRATED THAT BARRINGER Crater in Arizona and the craters of the Moon all had formed from impact events, Eugene Shoemaker reasoned that geology—or its lunar equivalent, selenology—would be the first field science that humans would conduct on another world. This led him to create an astrogeology program for the United States Geological Survey (USGS). As the program's chief scientist, Shoemaker brought the USGS into alliance with NASA, and suggested that geologists should travel to the Moon.

Selected in 1959, NASA's first cohort of astronauts—NASA Group 1, the "Mercury Seven"—consisted of military test pilots. Along with flight training, these men had engineering backgrounds. So important was engineering to spaceflight that Sergey Korolyov, chief designer for the Soviet space program, was beginning to take spacecraft engineers who were not military pilots and make cosmonauts of them.

NASA's second and third astronaut groups, selected in 1962 and 1963 respectively, consisted of test pilots similar to the first group, and this would be a good decision. In Project Apollo and Project Gemini, test-piloting skill and engineering would save astronauts from nearly fatal in-flight emergencies. Since Apollo astronauts were to land on the lunar surface, however, the scientist-astronaut concept did make its way into planning for NASA Group 4.

As the pilots of groups 2 and 3 entered astronaut training, Shoemaker was top among potential candidates for Group 4, until he was diagnosed with adrenal insufficiency. This disqualified him from the astronaut program, but he spearheaded geoscience for NASA's unpiloted Ranger probes, and for a new program called Surveyor, which was to soft-land lunar instruments to see what potential landing sites had in store for human missions.

Around 1964–65, Shoemaker also organized a handful of NASA geologists to provide astronauts with geology field training. Geologists in this program included David S. McKay (1936–2013), who, three decades later, would find evidence of possible microscopic fossils in a meteorite from Mars. This lined up McKay to help start NASA's Astrobiology Institute, where he mentored astrobiology trainees, including the author of this book.

SEE ALSO: New Understanding of Craters (1948–60), Beginnings of Lunar Science (1964), Lunar Prospector and Surface Ice (1998)

Geologist and astronomer Eugene Shoemaker poses with his model of a lunar landing, which mimicked the landscape that astronauts might find when they arrive.

IMPROVING SPACE CAPABILITIES

THE FINAL MISSION OF NASA'S PROJECT Mercury on May 16, 1963, returned astronaut Gordon Cooper (1927–2004) to Earth after 34 hours in space. The United States was now committed to the human Moon program, but more than 22 months would pass before another US astronaut would leave the atmosphere. During the intervening period, news highlighting a series of firsts in human spaceflight came out of the USSR. The first woman in space, Valentina Tereshkova, orbited in Vostok 6 just a month after Cooper's flight. Sometimes the Soviets would have two piloted craft in space simultaneously, and in October 1964 came the news of the three cosmonauts in one Voskhod capsule, though it had been designed to carry just two.

Soviet chief space designer Sergey Korolyov did not have to subject cosmonauts to the same dangerous stunt again, because he soon had a new boss, Leonid Brezhnev, the new First Secretary of the Communist Party. Thus, sending up the next Voskhod with a crew of two, Korolyov managed a feat that would be a prerequisite to any agency seeking to place humans on the Moon. On March 18, 1965, cosmonaut Alexey Leonov (b. 1934) performed the first "spacewalk," twelve minutes of what NASA would later call extravehicular activity (EVA). Unknown at the time was that Leonov almost hadn't made it back into the capsule. His suit had overinflated such that he could not get through the door.

NASA was going through the final checks for Gemini 3, the first piloted mission of the new spacecraft, while Leonov was performing his historic EVA, a procedure that NASA would need to master prior to sending humans to the Moon. American astronauts Virgil "Gus" Grissom (1926–67) and John Young (1930–2018) orbited Earth for nearly five hours on March 23, 1965. Along with maneuvers performed to change the shape and altitude of their orbit, Grissom and Young also used the Gemini's thruster system to orient the craft during reentry to create lift that could change the location of the landing point. Two and a half months later, astronaut Edward White (1930–67) would perform the first American EVA on Gemini 4.

SEE ALSO: Human Computers (1963), Apollo 1 Fire (1967), Descartes Highland (1972)

Edward White performs the first American EVA on June 3, 1965, remaining outside the Gemini 4 for 22 minutes. In his right hand, he carries a device that helps him maneuver. His visor reflects the umbilical and tether lines anchoring him to the spacecraft and providing life support.

LEARNING TO RENDEZVOUS AND DOCK

HUMAN LUNAR MISSIONS REQUIRE COURSE corrections, rendezvous and docking, and the ability to function in spaceflight for at least a week. Power was one challenge, since batteries could last just limited amounts of time, but Gemini V introduced the spacecraft fuel cell, which combined hydrogen and oxygen to produce electricity—and drinking water too. This provided Gemini V an eight-day space mission in August 1965, while Gemini VII flew in space for 14 days in December of the same year. Such extended spaceflights allowed scientists to start examining how the body, especially the nervous system, adapted to weightlessness. Whereas astronauts were known to become nauseous soon after reaching a weightless state, increasing mission times from hours to days on Gemini hinted that the nervous system could easily adapt to the sensations of the new environment.

Docking between spacecraft required practice on several missions. Plans called for Gemini astronauts to dock with a target vehicle called Agena. But while docked to an Agena in March 1966, astronauts Neil Armstrong (1930–2012) and David Scott (b. 1932) were thrown into a spin that worsened after they undocked, because of a problem in the altitude control system (ACS) of their Gemini VIII capsule. Rolling rapidly, the men were subjected to a G-force of 3.5 (they weighed 3.5 times their normal weight), plus the disorientation of the spin itself. They would have lost consciousness and died had not Armstrong turned off the ACS and countered the spin using thrusters from the reentry control system (RCS). The perfection of docking in later Gemini missions would enable Armstrong to become the first person to walk on the Moon, and Scott to be the first to drive on the Moon, and the first to drop a hammer and feather on the lunar surface.

Although the ACS problem was fixed after Armstrong and Scott returned to Earth, a radar malfunction almost prevented astronauts Jim Lovell (b. 1928) and Buzz Aldrin (b. 1930) from lining up with an Agena during Gemini XII. Holding a doctorate from MIT, his dissertation titled "Manned Orbital Rendezvous," Aldrin was nicknamed "Dr. Rendezvous." That name looked all the more appropriate when Aldrin pulled out a sextant, did the math, and guided Lovell's approach.

SEE ALSO: Telescopic Study of the Moon Begins (1609), Neutral Buoyancy (1966), One Giant Leap (1969), Extended Missions (1971)

A view of the Agena from a window of Gemini VIII, 1966.

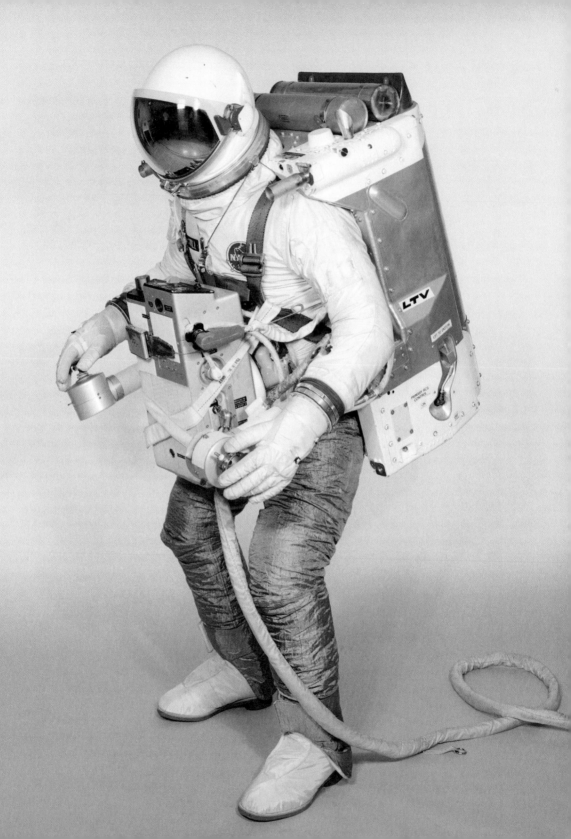

NEUTRAL BUOYANCY

Developed as a bridge between projects Mercury and Apollo, the Gemini program was like the middle child of a family who makes a series of remarkable achievements, sometimes with growing pains along the way. One major Gemini accomplishment that came with growing pains was learning how to perform extravehicular activity (EVA) that might be needed during a lunar flight. During Gemini 4, astronaut Edward White (1930–67) had exited the capsule for 22 minutes and tested a handheld maneuvering device before reentering the capsule without mishap. As with cosmonaut Alexey Leonov, however, Gemini soon found that movement was difficult in an inflated suit. Knowledge of Leonov's close call would have helped NASA, but the USSR had released only the good news that he'd completed an EVA.

After his EVA on Gemini IX, astronaut Gene Cernan (1934–2017) would characterize his suit as having "all the flexibility of a rusty suit of armor," a reality that prevented him from testing the astronaut maneuvering unit (AMU), a backpack with jets designed to help astronauts move through space. Working for two hours or more, Cernan, and then astronauts of Gemini X and Gemini XI, found that they fatigued quite easily, since it took a lot of energy to move in such a stiff suit. Stiffness in the gloves furthermore made it difficult to manipulate simple tools, and weightlessness made it challenging merely to remain anchored while pushing or twisting something.

Everything changed when astronaut Buzz Aldrin set out to train for the EVA in preparation for Gemini XII, scheduled for a November 1966 launch. A new training procedure was up and running, neutral buoyancy, in which the astronaut was immersed in a deep pool, wearing an EVA suit that was weighted appropriately to cancel out his buoyancy. Assisted by divers as he practiced, this arrangement made Aldrin effectively weightless. This allowed him to learn precisely how to perform everything that he would need to do during three EVAs totaling five and a half hours. By the end of the mission, Aldrin had demonstrated that an astronaut could do useful work in an EVA setting, a capability that was vital if humans were to visit the Moon.

SEE ALSO: Learning to Rendezvous and Dock (1965–66), One Giant Leap (1969)

A space suit developed for the Gemini program, 1966. The suit, known as the astronaut maneuvering unit, included a backpack containing a life-support system. The suit was ultimately never used.

TRAGEDIES

ALONGSIDE ADVANCES IN DOCKING, EXTRA-vehicular activity, and orbital maneuvering, NASA made a soft landing on the Moon with the Surveyor 1 probe touching down at Oceanus Procellarum, June 2, 1966, four months after the Soviet Luna 9 also had soft-landed. Later that year, NASA's Lunar Orbiter 1 became the first American probe to enter lunar orbit, a necessary step for verifying landing-site safety for more Surveyors, and for Apollo astronauts. But 1966 also was a year of Space Race tragedies.

In the USSR, after introducing cosmonauts to the new Soyuz spacecraft, chief designer Sergey Korolyov was rushed to the hospital. He died several days after surgery, possibly in connection with an abdominal tumor, or the anesthesia, although he had multiple health problems set off by his days as a gulag prisoner. In the course of two decades, Korolyov had spearheaded the first satellite, the first man in space, the first extravehicular activity, and most recently the *N1* Moon rocket. Conceived to launch humans on flyby missions of Venus or Mars, the future of the N1 would now rest upon Korolyov's deputy, Vasily Mishin (1917–2001), who was looking at the Moon as a possible first destination. Despite his genius, Korolyov was still unknown to anybody outside the Soviet government. Now, Soviet leader Leonid Brezhnev decided to announce Korolyov's identity to the world and remember the chief designer in a state funeral in Red Square.

Faced with the first piece of evidence that the Soviet space program might be in trouble, Americans wondered if they might finally be pulling ahead in the Space Race. They were at least catching up, but not without tragedies of their own. As with the Mercury program, all Gemini missions returned astronauts safely to Earth. But two Gemini astronauts—Elliot See (1927–66) and Charles Bassett (1931–66)—died when their T–38 Talon jet struck the McDonnell Aircraft building, where their Gemini 9 spacecraft was being prepared. They were not the first astronauts to perish in training aircraft, and they would not be the last. Nor would aircraft account for all astronaut fatal accidents during the Space Race.

SEE ALSO: Verne Inspires the Father of Astronautics (1870s), Sputnik (1957), First Pictures from the Moon's Farside (1959), Humans Enter Space (1961)

Sergei Korolyov, 1946.

APOLLO 1 FIRE

ASTRONAUT VIRGIL "GUS" GRISSOM HAD almost drowned when the hatch of his Mercury capsule had blown open too early after splashdown in July 1961. That same year, cosmonaut Valentin Bondarenko (1937–61) had burned to death in an isolation-chamber experiment, where the air had contained a high fraction of oxygen (O_2) compared with nitrogen (N_2) and technicians had been unable to open the hatch. The Soviets had kept the tragedy secret. Meanwhile, North American Aerospace, the contractor for Apollo, had designed an inward-opening hatch for the new spacecraft.

Apollo's life support system was to supply a cabin atmosphere of pure O_2 at 5 pounds per square inch (psi) (34.5 kilopascal [kPa]). This is higher than the usual 3.1 psi (21.4 kPa), the partial pressure of O_2 at Earth's surface, where it is mixed with N_2 and other gases, yielding a total pressure of 14.7 psi (101.4 kPa). North American had wanted to build Apollo with a dual N_2/O_2 atmosphere, and space experts had worried about potential fires during spaceflight. The potential for cabin fire on the launchpad hadn't come up, however, even though it was much more likely than an inflight fire, since prior to launch, the cabin needed an internal pressure of 16 psi (110.3 kPa). In such an environment, everything can burn, and that's what happened on January 27, 1967.

Grissom and his crewmates Edward White and Roger Chaffee (1935–67) were going through a launch rehearsal for mission AS-204, remembered as Apollo 1. It was a "plugs-out test"; the spacecraft and its launcher, the Saturn 1B rocket, would run on internal power through a full countdown. This was a block-I Apollo, the earliest piloted version of the spacecraft. The mission coming up in a few weeks was to be a shakedown flight in low Earth orbit, and the astronauts had breezed through the same rehearsal weeks before. This time, however, a spark went off in a wire behind a control panel. Fire enveloped the cabin within seconds; the high internal pressure against the inward opening hatch prevented escape; and the three astronauts lost their lives.

SEE ALSO: Improving Space Capabilities (1965), Neutral Buoyancy (1966), Re-engineering Apollo (1967)

The crew of the Apollo 1, photographed during a training in 1966, ten months before a deadly plugs-out test. From left to right: Virgil "Gus" Grissom, Edward White II, and Roger Chaffee.

REENGINEERING APOLLO

THE APOLLO 1 INVESTIGATION INVOLVED engineers, fire specialists, and a congressional committee, but NASA appointed astronaut Frank Borman (b. 1928) to lead the team. Investigators highlighted the spark in faulty wiring, causing cabin materials to ignite. They reported that the fire had spread rapidly because of the oxygen (O_2) at 16 psi (110.3 kPa), and that the inward-opening hatch had prevented escape. Testimonies painted a picture of a program that had failed to prognosticate a mundane ground accident in the rush to launch a block-I Apollo as part of an effort to beat the USSR in a manned mission to the Moon.

The year 1967 was not an optimistic time in America. The Vietnam War was escalating, plus there was overt resistance against the struggles for civil rights, which often ended up in violence. With the fate of NASA's Apollo program hanging in the balance, the USSR once again appeared to be winning in space, having delivered three probes into lunar orbit and soft-landed Luna 13 on the lunar surface. In America, the most positive-looking space program was not the real one, but the television series *Star Trek*, featuring the fictional twenty-third-century USS Enterprise crew that included an African woman, Asians, people speaking multiple languages, an extraterrestrial, and, as of season 2, even a Russian. Each week, *Star Trek* delivered stories about social justice, leading Dr. Martin Luther King Jr. (1929–68) to remark that it was one of the only television programs that he allowed his children to watch.

By April 1968, King was dead, and so was Yuri Gagarin, the first human in space, a victim of yet another training aircraft accident. Two months later, humanity lost Robert Kennedy (1925–68), brother of the US president who had initiated Project Apollo. Amidst all this, engineers redesigned the still-unfinished block-II Apollo with a new hatch, a protected electrical system, and fireproof materials. In spaceflight, the air would be 100 percent O_2 at 5 psi (34.5 kPa), as planned before, but a 60/40 oxygen/nitrogen (N_2) mixture was introduced for the ground phase.

SEE ALSO: Apollo 1 Fire (1967), Lunar Tortoises (1968), Reaching for the Moon (1968)

This photo shows damage to the exterior of the Apollo I, caused by the fire that killed its crew in 1967.

DECLARING PEACE ON THE MOON

The US AND SOVIET SPACE PROGRAMS were both in a hurry to send humans beyond low Earth orbit (LEO), and both made fatal errors. Within three months of the Apollo 1 fire, the USSR would have its own tragedy with the new Soyuz spacecraft. Launched to test the Soyuz in LEO, Soviet cosmonaut Vladimir Komarov (1927–67) suspected that the craft was not ready for human flight. After troubleshooting problem after problem in spaceflight and nursing the ailing craft, he survived reentry, flying manually against all odds. But then he crashed when the Soyuz parachutes failed. His last action was to condemn the management of a space program that would put such an unready craft on the launchpad. Following the usual Soviet pattern, Komarov's death did not go public.

Eight years would pass before Soviets and Americans would fly together in the Apollo-Soyuz Test Project, the first international space docking. Decades would pass until cosmonauts and astronauts would share a space station. However, 1967 did see one positive development in the international space arena. January 27, 1967, the same day as the Apollo 1 tragedy, the United Nations initiated the Treaty on Principles Governing the Activities of States in the Exploration and Use of Outer Space, Including the Moon and Other Celestial Bodies. Known today as the Outer Space Treaty, the agreement was signed immediately by the United States, the United Kingdom, and the USSR. As of July 2017, 107 countries had ratified the Treaty and another 23 had signed without actually ratifying.

Forming the basis of international space law, the treaty limited utilization of the Moon and other celestial bodies to peaceful purposes. No weapons of mass destruction or military bases would be allowed on the Moon, on other celestial bodies, or in space. No government could lay claim to the Moon or other celestial resources. In 1979, a new treaty proposed that the international community should have jurisdiction over the Moon and other celestial bodies, but no nation with human spaceflight capability had signed the new treaty as of January 2018.

SEE ALSO: Two's Company, Three's a Crowd (1964), Tragedies (1966), Apollo 1 Fire (1967)

The signing of the Outer Space Treaty in London, 1967. British foreign secretary George Brown (center) and Mikhail Smirnovsky (left), the Soviet ambassador to the UK, converse, as Phillip Haiser (right), representing the United States, signs the agreement.

LUNAR TORTOISES

N A SMITHSONIAN LECTURE IN 2008, ASTRO-
naut Frank Borman remembered the
Apollo 1 fire investigation that he had led
as a pivotal moment in NASA history. Not only
did the tragedy lead to design changes that
made the spacecraft much safer, but the investigation also exposed a plethora of problems in
the Moon program itself. This led to massive
systemic changes in management of spaceflight
and the development and testing of equipment.
Borman attributed many of the new administrative policies and their execution to Apollo
program manager George Low (1926–84).

Low's management approach penetrated all
through the human Moon program, but there
were disagreements on strategies for testing
space hardware. Wernher von Braun wanted
to launch each stage of the Saturn V on a separate test flight, but CIA analysis suggested that
the USSR was preparing to send cosmonauts
on a circumlunar trajectory, a flight around
the Moon. To save time, another high-ranking
NASA manager, George Mueller (1918–2015),
insisted on "all-up" testing of a fully assembled
Saturn V. All-up testing worked when the first
Saturn V launched in November 1967 as the

Apollo 4 mission. Ten weeks later, a Saturn 1B
carried a legless prototype lunar-excursion
module (LEM) into low Earth orbit, where
LEM engines and landing abort systems passed
flight-testing. But a second Saturn V, launched
in April 1968 as the unmanned Apollo 6 mission,
suffered severe "pogo" oscillations (vibrations
along the long axis of the rocket, like a pogo
stick) that would have killed astronauts, plus
two engines shut down on stage II, although the
craft compensated and achieved orbit.

In September 1968, the USSR did launch
"cosmonauts" on the Zond 5 mission around
the Moon and back to Earth, although not
human cosmonauts. They were tortoises, meal-
worms, and other creatures, but they did not
survive reentry through Earth's atmosphere,
due to a guidance program that exposed them
to twenty times the force of gravity. Animals
on Zond 6, two months later, also died, due to
a loss of cabin pressure. The Soviets were not
ready to send humans around the Moon, but
America did not know this.

SEE ALSO: Reengineering Apollo (1967), Reaching for
the Moon (1968)

A crane places an Apollo spaceship on top of a Saturn V rocket in preparation for the Apollo 4 mission. Launched in June 1967, this mission was an "all-up" test of the Saturn V that was conducted without first flight-testing each stage separately.

REACHING FOR THE MOON

POLLO 6, THE SECOND FLIGHT OF A Saturn V, ended with partial success; despite the pogo oscillations and engine shutdowns, the rocket had recovered by running the functional engines for extra time. Thus, while Wernher von Braun's team still had much work ahead, NASA was confident that it had an almost-ready launch vehicle. Even so, 1968 was a difficult year in America because of developments outside the space program. So powerful were the F1 engines of the Saturn V's first stage that the Apollo 6 launch, occurring on April 4, registered on earthquake sensors across North America. It would not register as more than a footnote on the Florida evening news, however, as the press was buried in President Lyndon Johnson's recent announcement that he would not seek reelection. Moreover, several hours after the launch, Martin Luther King Jr. was murdered in Tennessee.

Since the new block-II command module looked good by spring 1968, Apollo 7 was scheduled as a "C mission"—three astronauts in a command/service module (CSM) flight in low Earth orbit (LEO)—for October. The plan was for a subsequent crew to perform a "D mission" featuring a CSM and lunar module (LEM) in LEO. A third crew would then perform an "E mission," a CSM-LEM flight in elliptical medium Earth orbit, reaching 3,500 nautical miles (6,500 kilometers, or 4,028 statute miles) into space. But the suspected upcoming Soviet circumlunar flight, plus delays in preparation of the LEM, triggered a bolder plan. The crew that was training for the E mission was cycled forward. Apollo 9 became Apollo 8. Consisting of Frank Borman as commander, Jim Lovell, and William Anders (b. 1933), they would fly into orbit around the Moon, so long as Apollo 7 was successful—which it was, except for the fact that the astronauts developed colds and consequently refused to don their helmets before reentry, which got them into trouble with flight directors. Meanwhile, the Soviet Zond 5 and Zond 6 probes made their circumlunar journeys. Borman, Lovell, and Anders now prepared for what NASA Flight Operations Director Christopher Kraft would later characterize as the greatest mission of Project Apollo.

SEE ALSO: Lunar Tortoises (1968), Earthrise (1968), Dress Rehearsals (1969)

Apollo 7 crew members, from left to right: commander Walter M. Schirra Jr. (1923–2007), lunar module pilot Walter Cunningham (b. 1932), and command module pilot Donn F. Eisele (1930–87). The Apollo 4 launched from Cape Canaveral on November 9, 1967. The mission lasted nine hours and was deemed a successful test of the Saturn V launch vehicle.

EARTHRISE

1968

ASTRONAUT BILL ANDERS SNAPPED ONE of the most famous pictures in history on December 24, 1968, after the Apollo 8 spacecraft had come around from the far side of the Moon. Years later, when *Life* magazine chose the photo as one of the top hundred images of the twentieth century, Anders shrugged off the achievement as due to the fact that all three astronauts had taken a series of photos; he simply had grabbed a camera that happened to be loaded with color film and which had a particularly good telescopic lens.

After a year of upheaval in the US and throughout the world, humanity saw Earth from a new perspective. At an average distance of 207,600 nautical miles (384,475 kilometers, or 238,900 statute miles), Earth, the planet that had cradled civilization since the days when humans had worshipped the Moon as a god, fit into a single frame surrounded by darkness. It was a swirly blue-and-white marble from this vantage point, what astronomer Carl Sagan would later call a pale blue dot. And yet all humanity was living on it—except for mission commander Frank Borman, command-module pilot Jim Lovell, and Anders, who served as "lunar module pilot" although there was no lunar module on this mission.

For ten orbits, stretching over a 20-hour period, the astronauts flew 60 nautical miles (110 kilometers, or 69 statute miles) above the lunar surface, scoping out landing sites for future missions, studying geological features, and mapping gravity changes resulting from the rugged topography and uneven densities within the lunar crust. In mapping the gravity changes, the astronauts revealed previously unknown masses within certain craters. Astronaut geology training was finally coming into play, but there were also some lighter moments: learning that a particular triangular form near the Sea of Tranquility was unnamed, Lovell named it Mount Marilyn, after his wife.

On the tenth orbit, the craft's main engine fired, returning Apollo 8 to Earth on December 27. En route, Lovell practiced a manual course-correction procedure that would help save his life sixteen months later on Apollo 13. Upheaval had defined 1968, but *Time* magazine chose Borman, Lovell, and Anders as Men of the Year.

SEE ALSO: Reengineering Apollo (1967), Lunar Tortoises (1968), Reaching for the Moon (1968)

The Earth rises into the Moon's sky in this photo taken from Apollo 8. Astronauts Frank Borman (b. 1928), James Lovell (b. 1928), and William Anders (b. 1933), who took this photo, were the first humans to view Earth from this perspective.

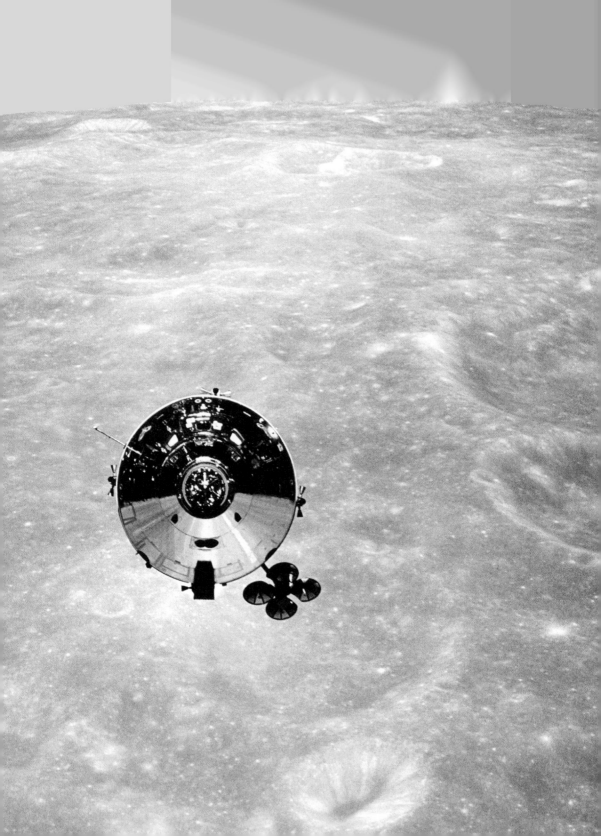

DRESS REHEARSALS

WHEN APOLLO 8 RETURNED TO EARTH, astronauts had yet to fly a lunar module (LEM) in space and test rendezvous and docking between the LEM and command/service modules (CSM). They also had yet to test the backpack-shaped Portable Life Support System (PLSS) developed to enable extravehicular activity (EVA) on the Moon without the burden of umbilical lines. Developed by Grumman Aircraft Engineering Corporation on Long Island, New York, the LEM had passed testing on the unpiloted Apollo 5 mission, but was behind schedule; this is why Apollo 8 had been recast as a CSM-only mission to the Moon. The LEM was still too heavy for a lunar landing, but all other LEM capabilities could be tested in the low Earth orbit (LEO) "D-mission" setting of Apollo 9. Orbiting ten days in March 1969, Apollo 9 took James McDivitt (b. 1929), David Scott, and Russell Schweickart (b. 1935) through a mission, highlighted not only by maneuvers between the two craft that would be vital to lunar missions, but also by Schweickart performing an EVA with the PLSS. A bout of space adaptation syndrome (nausea and vomiting caused by weightlessness) prevented Schweickart from performing additional EVAs, but the stage was adequately set for the next mission, Apollo 10.

Reaching LEO on May 18, 1969, after the S-IVB stage of their Saturn V rocket had completed its first engine burn, astronauts Thomas Stafford (b. 1930), John Young, and Gene Cernan checked out all systems before receiving word that they were "go for TLI." That meant trans-lunar insertion, reignition of the S-IVB engine that would send them to the Moon. Designated by the call sign "Charlie Brown," the CSM connected nose-to-nose with the LEM, designated as "Snoopy," and the two craft coasted toward the Moon until the CSM main engine fired to slow them into lunar orbit. Like the previous LEM, "Snoopy" was too heavy for a lunar landing; Grumman still had more work to do. But on May 22, Stafford and Cernan flew Snoopy down to 8.4 nautical miles (15.6 kilometers, or 9.67 statute miles) above Mare Tranquillitatis to assess the region where the next LEM crew would land in July.

SEE ALSO: Reaching for the Moon (1968), Earthrise (1968)

A photo of the CSM Charlie Brown taken from the LEM Snoopy. During the mission, commander Tom Stafford and lunar module pilot Gene Cernan flew Snoopy down to just 8.4 nautical miles (9.67 statute miles, or 15.6 kilometers) above Mare Tranquillitatis, the landing site for NASA's next mission.

ONE GIANT LEAP

ON JULY 16, 1969, NEIL ARMSTRONG, BUZZ Aldrin, and Michael Collins (b. 1930) launched from Kennedy Space Center and entered lunar orbit three days later. The US now had a comfortable lead in the Space Race, but didn't realize yet just how comfortable. On July 3, a Soviet N1 rocket on which lunar missions depended had exploded during a test launch. The Soviets would continue developing the N1 through 1972, but human lunar exploration was now an American project.

Soviet television did not broadcast the events that were about to unfold, as Armstrong and Aldrin separated the lunar module Eagle from the command module Columbia on July 20 and descended into the Mare Tranquillitatis. Approaching the lunar surface, Armstrong and Aldrin saw that the computer was taking them to a boulder field. Armstrong took manual control and searched for a new landing site, as Aldrin read off distances and altitudes. Seconds of fuel remained just before Armstrong set the craft on the ground.

Apollo 11 was the final engineering flight of the program. The prime mission was to land on the Moon, launch back into lunar orbit, and return to Earth in Columbia. Humanity remembers the mission best for Armstrong's words "That's one small step for man, one giant leap for mankind," uttered as he stepped onto the Moon, and for Armstrong's iconic photo of Aldrin with his visor closed. But the astronauts devoted most of their two and a half hours of extravehicular activity on the lunar surface to science. After gathering contingency samples— rocks and dirt that they could take to Earth, in case they needed to make an emergency departure—they collected samples more deliberately, including some in core tubes, and made careful rock sample selections. Seeing large gaps between the rocks in a container, Armstrong, on his own initiative, poured in dirt to fill the gaps. The astronauts also set up a prototype suite of instruments known as the Early Apollo Scientific Experiments Package, plus a Laser Ranging Reflector, designed to reflect light back precisely to its source, regardless of the angle at which it arrives.

SEE ALSO: Learning to Rendezvous and Dock (1965–66), Return to the Moon (1971)

Buzz Aldrin, photographed by fellow astronaut Neil Armstrong, walks on the surface of the Moon, leaving footprints in his wake.

BEGINNINGS OF
LUNAR FIELD SCIENCE

EQUIPPED WITH IMPROVED LANDING radar, Apollo 12 astronauts Charles "Pete" Conrad (1930–99) and Alan Bean (1932–2018) aimed for accuracy as they entered lunar module Intrepid on November 19, 1969. They were headed for NASA's Surveyor 3 probe, which had soft-landed in 1967, in Oceanus Procellarum, 2,000 kilometers (1,240 miles) from the Eagle landing site.

The astronauts set up the Apollo Lunar Surface Experiments Package (ALSEP). An expanded, more complex version of the instrument package at the Eagle landing site, ALSEP included seismometers to measure moonquakes (shaking of the lunar surface), devices for producing mini-moonquakes, magnetometers, and instruments to measure charged particles, the lunar "atmosphere," particles ejected from meteorite impacts, gravity fluxes, solar wind, and heat flow near the lunar surface, all powered by a nuclear electric generator.

Along with deploying the ALSEP, Conrad and Bean were tasked with collecting geological samples from lunar regolith, the mesh of inorganic dirt and rocks that covers the lunar bedrock. Like the Apollo 11 landing crew, Conrad and Bean were exploring a mare, a dark lowland, which was expected to harbor rocks that had

once been part of the mare bedrock plus rocks that had been ejected from the deep crust by crater-forming impacts. In 1969, the idea that mare surfaces had formed from basaltic lava flooding impact basins and craters after rising up through cracks in the crust was a hypothesis. The hypothesis was confirmed, based on Basalt samples collected during Apollo 12 and other lowland missions.

Having landed within walking distance of Surveyor 3, as planned, Conrad and Bean retrieved parts from the probe, including the television camera. Inside the camera, biologists found *Streptococcus mitis*. Most likely, it came from somebody's mouth after the mission, but native lunar life forms could not be ruled out. Thus, astronauts of the first three lunar landing missions wore respirators while transferring from spacecraft to an isolation chamber before entering a quarantine facility for 21 days. Apollo 11 astronauts also exited their spacecraft in biohazard suits, but these were abandoned beginning with Apollo 12.

SEE ALSO: Peak of Lunar Volcanic Activity (3.8–3.5 Billion Years Ago), Making Moonquakes (1969), The Lunar Receiving Laboratory (1969), Elucidating Lunar History (1970s–1980s)

Pete Conrad takes a look at the television camera on the Surveyor 3. The lunar module that carried Conrad and Alan Bean to the Moon is visible in the distance.

MAKING MOONQUAKES

WHEN AN OBJECT IMPACTS THE MOON, it generates waves, some of which can travel directly to any seismometer deployed on the lunar surface. At the same time, if there is a high-density structure within the Moon such as an iron-rich core, some waves from the impact will bounce from it. This will stimulate the seismometer differently, as will other density interfaces, such as interfaces between crust and mantle. Utilizing this phenomenon, scientists could learn something about the Moon's inner structure. Furthermore, increasing the number of seismic stations across the Moon could reveal more detail.

When Apollo 12 astronauts deployed the first full Apollo Lunar Surface Experiment Package (ALSEP) in November 1969, the Apollo 11 package had long since stopped operating. But the full ALSEP, including the seismic instrument, was designed for long-term operation. To create moonquakes, ALSEP included a mortar and a ground thumper that would be tested on later missions, but there was also another way to shake the Moon. Like all lunar modules, Apollo 12's Intrepid consisted of two stages: a descent stage to land the astronauts softly, plus an ascent stage to carry them back into lunar orbit. Once the two Moon-walking astronauts were back in the command module with the command module pilot, Intrepid was jettisoned to crash into the lunar surface. The Intrepid impact registered on the ALSEP seismic instrument, so mission designers planned to impact the Moon with future LEM ascent stages, plus a booster that could make a bigger impact, namely the S-IVB stage of the Saturn V.

As a third Apollo landing mission geared up for an April 1970 launch, there was plenty of work to do with the geological samples that the first two crews had brought to Earth. Biologists were investigating whether the lunar materials harbored disease-causing organisms. Other scientists were studying the physical properties of lunar dust particles, which were so tiny, sharp, and electrically charged that they were eating away at spacesuits, such that joints on the spacesuits were not operating correctly and some suits were developing air leaks. Astronauts were also breathing the dust into their lungs.

SEE ALSO: Peak of Lunar Volcanic Activity (3.8–3.5 Billion Years Ago), The Lunar Receiving Laboratory (1969), Return to the Moon (1971), Elucidating Lunar History (1970s–80s)

A NASA photographic technician came into contact with lunar dust when handling film magazines used by the crew of Apollo 11.

THE LUNAR
RECEIVING LABORATORY

ASTRONAUTS OF APOLLO 11 COLLECTED 22 kilograms (48.5 pounds) of lunar material, whereas Apollo 12 returned to Earth with 34 kilograms (75 pounds). Lunar samples entered the Lunar Receiving Laboratory (LRL), located at NASA's Johnson Space Center (JSC), Building 37, the same building where the astronauts spent their quarantine.

Since Apollo 11 was the first space mission to carry lunar samples to Earth, there was no way to rule out the possibility that the samples might contain lunar organisms, including pathogens, before the Apollo 11 astronauts had come home. But scientists began testing lunar materials from the first two landing missions on different organisms. These included plants such as tobacco and conifer seedlings, invertebrates such as shrimp and insects, protists such as paramecia, and vertebrates such as Japanese quail and germ-free mice.

No pathogens turned up, plus lunar materials were deficient in water, so scientists concluded that the Moon had no native life forms. This meant that the quarantine would not be required for returning lunar astronauts beginning with Apollo 15, but it did not mean that microorganisms from Earth could not survive in the lunar environment.

Scientists identified basic categories of lunar materials. Pulverized regolith consisted of glasses with crystals and angled rock fragments. Lunar volcanic rocks were basalts with high levels of plagioclase feldspar minerals, accompanied by clinopyroxene and another type of mineral called ilmenite. There were also breccias, formed when impacts had compressed volcanic rock fragments and finer regolith. A breccia and some regolith samples from Apollo 12 featured an unusual combination of elements known by the acronym KREEP (potassium [K], Rare Earth Elements, Phosphorus). Appearing also in Apollo 14 and 15 samples, KREEP would help demonstrate that the early Moon had gone through a molten state, during which lighter elements had floated upward. Meanwhile, the samples brought to Earth on the first two landing missions were enough to reveal impacts and volcanism as prime forces shaping the early Moon and all inner planets. Since some lunar rocks dated back to a time that geological forces had erased on Earth, Apollo lunar samples were on course to revolutionize Earth science.

SEE ALSO: Beginnings of Lunar Field Science (1969), Preparing for New Missions (2018)

A lunar sample collected by Neil Armstrong and Buzz Aldrin, as they performed their EVA on the Moon. It was determined that this rock was rich in magnesium, similar to some rocks on Earth.

A SUCCESSFUL FAILURE

BY 1970, THE AMERICAN PUBLIC SAW MOON exploration as routine, even though NASA planned a fascinating itinerary for the next mission, Apollo 13. The landing site, Fra Mauro highlands, might consist of lunar bedrock ejected billions of years ago by the massive impact that had carved the Imbrian Basin. But with the Vietnam War and draft protests dominating the news and many people's personal lives, and with an announcement on April 10 that the Beatles were breaking up, lunar science was not a media priority.

Even worse, reporters who did attend NASA press conferences asked questions that suggested that the American public had a case of triskaidekaphobia—fear of the number 13. Unlike certain hotels, NASA did not have its elevators skip the thirteenth floor. Quite the contrary, the Agency actively opposed superstition. During spaceflights, Mission Control regularly poked fun at astrology by issuing "horoscopes" for crewmembers based on their daily work assignments. The opposition to superstition persisted down to Apollo 13 mission commander Jim Lovell and flight director Gene Kranz (b. 1933). To make his point, Kranz scheduled the liftoff for 13:13 hours CST.

At 55 hours and 55 minutes into the flight, a flawed oxygen tank exploded, crippling the command/service module *Odyssey*, changing the mission to a struggle to return Lovell and his crewmates Fred Haise (b. 1933) and Jack Swigert (1931–82) to Earth alive. This required utilizing the lunar module Aquarius, a craft designed to sustain two people for two days, to sustain three people for four days, during which time Haise became febrile from a urinary tract infection. This task required manual course corrections, and a command module reboot on battery power to enable reentry, all this with no certainty as to whether the heatshield and parachute systems were still healthy enough to do their jobs. After four days of peril, the astronauts did make it home, because everything that had to go right actually did. Some might say that the astronauts got "lucky," but their training and the ingenuity and refusal of controllers and staff on Earth to accept failure had coalesced into an episode that Kranz would characterize as NASA's "finest hour."

SEE ALSO: A Lunar Facelift (3.9–3.1 Billion Years Ago), A New Model for Lunar Motion (13th Century), Earthrise (1968), Return to the Moon (1971)

A view of the damaged Apollo 13. The explosion of oxygen tank number 2 blew away one of the panels (right) and severely damaged the spacecraft's service module. Because the accident happened on the way to the Moon, the crew was able to utilize the lunar module Aquarius as a lifeboat.

RETURN TO THE MOON

PRIOR TO THE IN-FLIGHT EMERGENCY THAT led to a mission abort, Apollo 13 sent its SIV-B stage on a collision course with the Moon. The SIV-B hit on April 14, 1970, close to the site where Apollo 12 astronauts had landed, renamed Mare Cognitum, "the Sea that has become known." The collision generated seismic signals that registered at the Apollo Lunar Surface Experiments Package (ALSEP) at the Apollo 12 landing site. The collision also produced a 30-meter impact crater that would be photographed, along with the smashed SIV-B, by NASA's Lunar Reconnaissance Orbiter in 2010.

Launched January 31, 1971, with the goal of exploring Fra Mauro, the site originally intended for Apollo 13, the crew of Apollo 14 sent its own SIV-B stage on a lunar collision course and planned to do likewise with the ascent stage of lunar module (LEM) Antares. Throughout the remainder of Project Apollo, astronauts would leave more ALSEPs on the lunar surface, while poundings from SIVBs, LEM ascent stages, and thumpers and mortars at ALSEP stations themselves would stimulate ALSEP seismometers, supplying data on the physical structure of the Moon.

During their 33.5-hour stay at Fra Mauro, astronauts Alan Shepard and Edgar Mitchell (1930–2016) also deployed a Laser Ranging Reflector (LRR) identical to the LRR that Neil Armstrong and Buzz Aldrin had set up at Mare Tranquillitatis. These LRRs, plus one more that Apollo 15 astronauts would deploy, would enable laser measurement of Earth's axis wobble and continental drift and calculation of the rate of the Moon's tidally generated movement away from Earth at four centimeters per year.

Apollo 14 returned useful geological specimens, the first gathered from a highland site. Dating back more than 4.5 billion years, Fra Mauro rocks have given support to the idea that the Moon is almost as old as the Solar System. Additionally, since basalts gathered from mare landing sites range only from 3.2 to 3.9 billion years in age, the Fra Mauro samples have confirmed a pre-Apollo hypothesis that lunar highlands are much older than the maria, with implications regarding the Moon's origin.

SEE ALSO Formation of the Moon (4.5 Billion Years Ago), One Giant Leap (1969), A Successful Failure (1970)

A photo of the Apollo 14 lunar module, 1971.

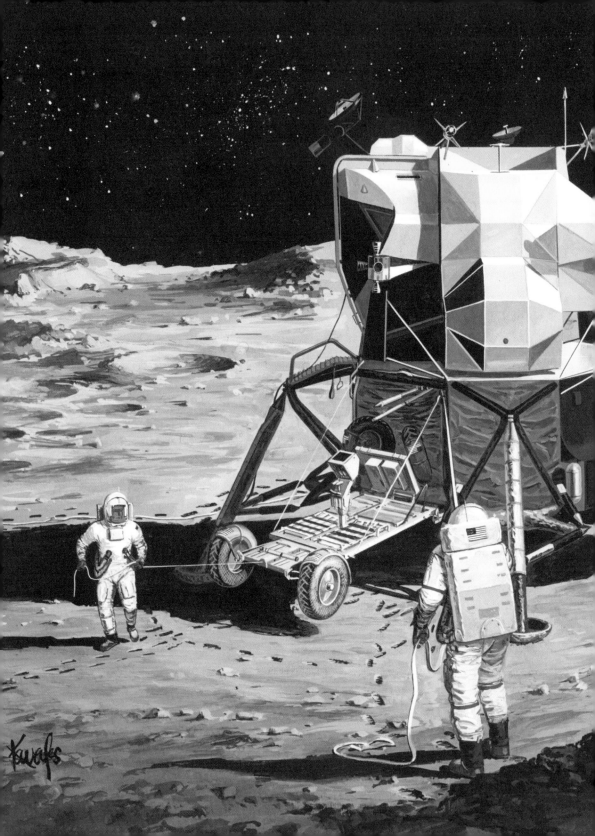

EXTENDED MISSIONS

APOLLO 15 WAS THE FIRST OF NASA's "J Missions," featuring longer stays on the Moon and increased mobility and science. Landing was in the Hadley Rille/Apennine region. In forming, the Imbrian Basin had pushed up the Appenine Mountains, which might hold samples of primal lunar crust. Northward of previous landing sites, this site required special maneuvers from the command/service module (CSM) Endeavour to get the lunar module Falcon to descend over the Apennine Mountains with enough fuel to land. The new maneuvers also enabled Falcon to carry a buggy called the Lunar Roving Vehicle (LRV) and supplies to support David Scott and James Irwin (1930–91) for three days.

The astronauts set up an ALSEP instrument suite and collected 78 kilograms (172 pounds) of samples, including anorthosite—white rocks comprised mostly of a type of plagioclase feldspar called anorthite. Relatively lightweight, anorthite had floated above heavier minerals in the Moon's molten era. These white rocks were indeed primal crust, the building material of lunar highlands.

Examining Hadley Rille, 300 meters (984 feet) deep and 1.5 kilometers (0.93 miles) wide, the astronauts saw that it appeared volcanic, the remnant of a lava tube whose roof had collapsed. At one point, Scott delivered a physics lesson on television by dropping a hammer and a falcon feather, which, as predicted, hit the ground simultaneously. He also planted a memorial plaque with names of fourteen astronauts—and cosmonauts—who had died in the service of space exploration.

Apollo 15 packed in a wealth of science, not just on the lunar surface, but also in orbit, where command module pilot (CMP) Alfred Worden (b. 1932) ran a suite of experiments, but the expedition was not free of glitches. Trying to pull out a drill that had jammed in the ground, Irwin suffered a cardiac arrhythmia, which at the time was suspected to be the result of hypokalemia—low potassium. Flight surgeons prescribed potassium for the next mission, Apollo 16. This led Apollo 16 commander John Young to complain about a side effect of potassium—flatulence—in somewhat coarse language, not realizing that his comments were being broadcast live on TV.

SEE ALSO: Telescopic Study of the Moon Begins (1609), Learning to Rendezvous and Dock (1965–66), The Lunar Receiving Laboratory (1969), Elucidating Lunar History (1970s–80s), Preparing for New Missions (2018)

In this illustration, the crew of Apollo 15 deploys the LRV. James Irwin (left) unfolds it using cables, while David Scott (right) lowers it to the lunar surface.

DESCARTES HIGHLAND

"There is nothing so removed from us to be beyond our reach or so hidden that we cannot discover."

—René Descartes, quoted on a NASA blackboard by lunar geologist Farouk El Baz, April 1972

B EING A J MISSION, APOLLO 16 BOASTED AN ambitious science itinerary and had its share of ups and downs. During the deployment of the ALSEP instrument suite, a cable to the heat-flow experiment dislodged after snagging on the leg of mission commander John Young; the experiment had to be abandoned. Furthermore, due to problems with pre-mission orbital photography, the geological features at the landing site did not all correspond to pre-mission expectations. Young and lunar module pilot Charles Duke (b. 1935) had to improvise. With one thousand hours of geology training (effectively the equivalent of a geology masters degree) under their belts, they returned an exquisite sampling of geological specimens, including some of the oldest rocks collected during the Apollo program. Most of the samples were breccias, which seemed to be all over the Descartes region, a highland southwest of Mare Tranquillitatis. This find disproved the notion that the region was covered by volcanic plains, while supporting the idea that the Moon, and all inner planets, had been subjected to frequent, massive impacts.

The Apollo 16 geological success paved the way for the next mission, which was to include an actual geologist. Selected into NASA Group 4, "the scientists," Harrison "Jack" Schmitt (b. 1935) held a PhD in geology from Harvard, and was lobbying NASA with a fascinating mission idea. As a landing site for Apollo 17, Schmitt proposed Tsiolkovsky crater. Named for the Russian astronautics pioneer, Tsiolkovsky was an enigma, being a mare on the far side, which consists mostly of highlands. It might thus be key to the puzzle of the Moon's asymmetry, but such an expedition would be risky, even with a satellite positioned to relay communications between the landing party and Mission Control. NASA rejected Schmitt's idea, but come December he was headed to Taurus-Littrow, a slim valley in the Montes Taurus Mountains.

SEE ALSO: Extended Missions (1971), Mission to Taurus-Littrow (1972), Apollo Biostack (1972), Cancelled Apollo Missions (1972–74)

Charles Duke (left) and John Young (right) sit on a Lunar Roving Vehicle designed for training.

MISSION TO TAURUS-LITTROW

1972

UNNING WITHIN THE MONTES TAURUS Mountains along the southeast edge of Mare Serenitatis is Taurus-Littrow, a narrow, deep valley, where Gene Cernan, mission commander of Apollo 17, landed the lunar module Challenger on December 11, 1972. Earlier that day, Cernan and lunar module pilot (LMP) Jack Schmitt had been orbiting above, where command module pilot Ronald Evans (1933–90) now flew solo, overseeing experiments covering lunar surface photography, astronomy, and the biological effects of space radiation. During the trip down, it had been clear that Cernan was in command, piloting Challenger with assistance from the LMP. Once outside the Challenger, however, the dynamic between the two men switched. Being a trained geologist, Schmitt was clearly in charge of the geological exploration, while commander Cernan took on a copilot role on the ground, applying his astronaut geology training to assist Schmitt, just as Schmitt had applied his flight training to assist Cernan. This was a new type of organizational structure for a space mission, but it made sense, and it yielded Apollo 17 the greatest scientific output of all six Apollo Moon landings. In all likelihood, the idea of landing parties consisting of a pilot with secondary training in field science and one or more field scientists with secondary training in spacecraft operations will be the model for future lunar and planetary exploration.

Taurus-Littrow was a geological treasure trove, but one finding that stands out was a scattering of orange and black glass beads. These were pyroclastic ash, fragments of volcanic material spewed from deep within the Moon billions of years ago. A titanium-containing mineral accounted for the orange color, but the beads also included crystals of a magnesium-iron silicate mineral called olivine. Olivine is very common in lunar rocks, so initially it was not the most intriguing aspect of the beads. In 2008, however, geochemists demonstrated that the olivine crystals enclosed tiny amounts of a familiar volatile compound: water. This has complicated the giant impact hypothesis of the Moon's origin, which predicts that volatiles should be absent from any materials that were part of the nascent Moon.

SEE ALSO: Apollo Biostack (1972), Cancelled Apollo Missions (1972–74), Elucidating Lunar History (1970s–80s), Preparing for New Missions (2018)

A close-up of lunar particles brought back by the crew of Apollo 17. These have been the finest particles ever observed, ranging only 20–45 microns (or 0.03 mm). In 2008, geochemists demonstrated tiny amounts of water within a mineral in the beads, complicating the giant impact hypothesis of the Moon's origin.

APOLLO BIOSTACK

1972

SINCE THE DISCOVERY OF THE VAN ALLEN radiation belts in the late 1950s, NASA had known that using the most direct trajectory to the Moon would expose astronauts to potentially lethal doses of radiation—protons and heavy ions, also called HZE particles, that have been trapped by the geomagnetosphere. Considering fuel, the inclination of the Moon's orbit around Earth, shielding capability of the Apollo hull, and the geometry of the belts, the solution was a trajectory that traversed only the corner of the inner belt, very rapidly, avoiding its most lethal radiation entirely, and that took astronauts through a fairly narrow region of the outer belt for just a few hours. This minimized exposure to trapped radiation, but did not eliminate it. Furthermore, outside the belts, space is full of un-trapped HZE particles, which exist as components of two types of deep-space radiation. One type, called solar particle events (SPEs), produces many low-energy HZEs periodically. The other type, called galactic cosmic radiation (GCR), includes smaller numbers of HZEs, but they are highly energetic and always present in the space between the outer Van Allen Belt and the Moon. It was unknown to what degree HZE particles from the outer belt and from GCR would affect life forms.

To study the issue, European scientists sent Biostack I and Biostack II, experiments respectively in the Apollo 16 and Apollo 17 command modules. Researchers measured HZE exposure in numerous biological species, including *Artemia salina* shrimp eggs, spores of *Bacillus subtilis* bacteria, and *Arabidopsis thaliana* plant seeds. HZE particles did not harm *B. subtilis* spores, nor did *Arabidopsis* seeds fare worse than control seeds on the ground, although the shrimp eggs exposed to HZEs in space proved more sensitive than the other organisms. Since Apollo 17, very few biological experiments have even flown outside the Van Allen belts. As for humans in deep space, there are very few data. Studies hint that there are reasons to be concerned about flights beyond LEO possibly elevating risks for cancer and cardiovascular conditions, cataracts, and other long-term effects, but the issue requires more study and we may not know the limits of human radiation tolerance until we return to the Moon and establish bases.

SEE ALSO: A New Discovery and a New Agency (1958–59), Beginnings of Lunar Science (1964), Descartes Highland (1972), Mission to Taurus-Littrow (1972)

An artist's rendering of three Van Allen belts. Lunar trajectories were designed to minimize exposure to the radiation of the belts. To avoid the most lethal radiation, astronauts very rapidly traversed only the corner of the inner belt and then passed through a fairly narrow region of the outer belt for just a few hours.

SERVICE MODULE

COMMAND MODULE

DOCKING ADAPTER

INSTRUMENT UNIT

ORBITAL WORKSHOP

SATURN APOLLO APPLICATIONS
CLUSTER CONFIGURATION

CANCELLED APOLLO MISSIONS

SPACE STATIONS, LUNAR BASES, ORBITAL telescopes, a piloted Venus flyby—these missions were on the drawing board of NASA's Apollo Applications Program (AP). Combining Saturn and Apollo components in new ways, AP would have introduced a progression of advanced command/service modules—block-III, block-IV, and block-V—subsequent to the geology-heavy Apollo "J missions." During the presidency of Lyndon Johnson (1908–73), NASA contracted the production of fifteen Saturn V rockets. This was enough to support lunar landings through Apollo 20, the last three missions envisioned to include landings inside the Copernicus and Tycho craters, or landings at the crater rim, followed by astronauts rappelling to the crater floor.

Apollo 20 was cancelled in January 1970 so that NASA could use the fifteenth Saturn V to launch the Skylab space station, the only AP project to survive budget cuts. While the US Congress controlled budgets, the president nevertheless had a great deal of influence through the veto power and a tradition of leading national policy through initiatives and political dealings. This put President Richard Nixon (1913–94) in a position to end America's first human lunar-exploration program.

Whereas the Johnson administration had overseen NASA cuts that would prevent lunar bases and the piloted Venus flyby from happening in the twentieth century, Johnson, like Americans overall, had supported Kennedy's Moon-landing initiative as a memorial to the fallen president, and later as a memorial to the crew of Apollo 1. Nixon, in sharp contrast, was eager to bring the lunar exploration era to a close, perhaps because he was still sore about losing the presidency to Kennedy in the 1960 election. Having no interest in what the astronauts were doing on the Moon, or what the lunar samples could reveal about the history of the Moon and Earth, Nixon attempted to get the Apollo 16 and Apollo 17 missions canceled in August 1971. Those two missions did survive, but Apollo 18 and Apollo 19 succumbed to cancellations made in September 1970. As for the Saturn V rockets that would have launched those two missions, today they are museum pieces at NASA centers.

SEE ALSO: Extended Missions (1971), Descartes Highland (1972), Mission to Taurus-Littrow (1972)

A model of the Skylab. Launched in 1973, it was the first American space station and featured a workshop and solar observatory.

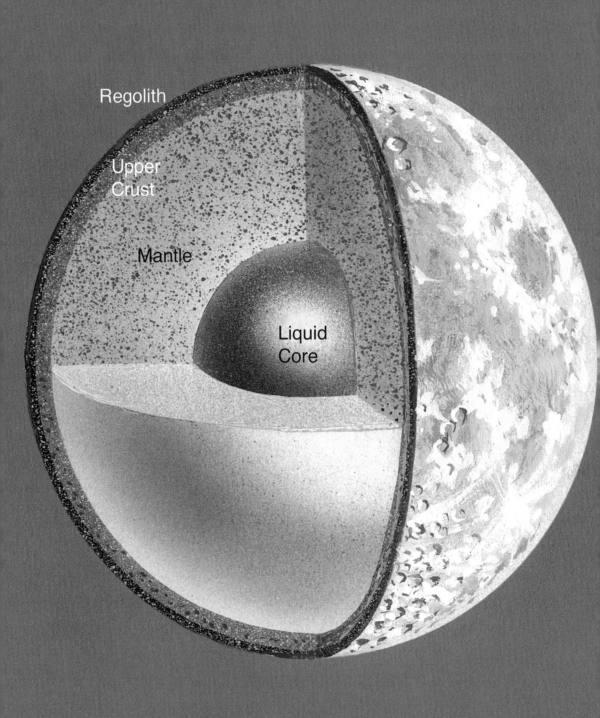

Regolith

Upper
Crust

Mantle

Liquid
Core

ELUCIDATING LUNAR HISTORY

A S NASA FOCUSED ON HUMAN LUNAR exploration in the 1970s, Soviet probes Luna 16, Luna 20, and Luna 24 delivered 326 grams (11.5 ounces) of lunar regolith in a robotic process that China's Chang'e 5 probe is scheduled to revive in 2019. Adding to these samples, scientists have since identified more than 300 meteorites as having lunar origins. Finally, there are the Apollo samples. Amounting to 382 kilograms (842 pounds), these specimens have been the most useful lunar samples.

By the mid-1970s, features of the Apollo samples suggested that the lunar crust had been molten for an extended time after the Moon's formation. Seismic measurements and other studies at Apollo landing sites furthermore revealed layering beneath the lunar surface. Atop the solid bedrock lay cracked but un-displaced rock, topped by boulders and smaller rocks, covered in turn with an outer layer of pulverized rock comprising the surface regolith. Back during the molten period, heavy minerals (pyroxene and olivine) sank deep, while lighter minerals (feldspar) floated. Consequently, a type of plagioclase feldspar called anorthite is abundant in highlands, such as the Apollo 15 and 16 landing sites.

After the crust hardened, big impacts carved basins and craters and released iron-rich basaltic magma in places where the crust cracked all the way through. Flowing lava spread over mare surfaces before hardening, giving maria their familiar darkened appearance.

Together with the Moon's large size compared with Earth, aspects of the Moon's orbit, and Earth's rotation, Apollo samples told a story that by the 1980s looked inconsistent with earlier hypotheses of lunar origins. Gradually more appealing was a scenario, introduced by William Hartmann and Donald Davis in 1975, that Earth had crashed with a Mars-sized planet not long after both worlds had coalesced and separated into crust, mantle, and core. In the process, Earth received most of the iron from the cores of both worlds and a rapid spin, while the Moon formed largely from mantle and crust material. In recent years, the model has been tweaked to account for an increasing number of finds, including water in glass beads discovered by the Apollo 15 and 17 astronauts.

SEE ALSO: Formation of the Moon (4.5 Billion Years Ago), Scientists Consider Lunar Origins (1873–1909), Extended Missions (1971), Descartes Highland (1972), Mission to Taurus-Littrow (1972)

The Moon's internal structure is comprised of three layers: the crust, mantle, and core. The crust encompasses both the regolith and upper crust shown in this diagram.

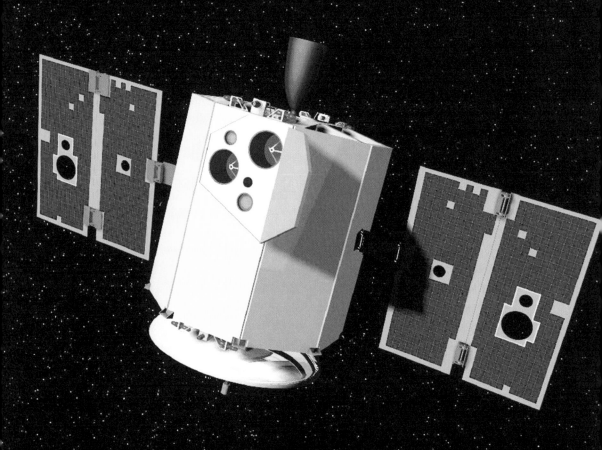

STUDYING LUNAR RESOURCES

PROJECT APOLLO TAUGHT SCIENTISTS THAT the lunar crust is made of various forms of silicate, consisting of silicon and oxygen. Also abundant are iron, magnesium, titanium, calcium, aluminum, and manganese. Scientists recognized that easy access to metals, minerals, and oxygen on the lunar surface could support heavy industry on the Moon.

By the 1980s, researchers had learned that the solar wind enriches the lunar regolith with helium-3, an isotope of helium that might serve as an advanced fuel for nuclear fusion power. Although fusion using only hydrogen isotopes will probably be achieved first, helium-3 fusion would offer the advantage of not producing neutrons. This would enable relatively small, lightweight fusion engines that could propel spaceships around the solar system very quickly. Helium-3, which is extremely rare on Earth, also has potential uses in nuclear medicine.

Also related to energy, lunar silicate can be processed into glass and electronics for production of photovoltaic cells (PVCs). In the wake of Apollo, physicist David Criswell realized that lunar-based solar power (LSP) could be beamed to Earth as microwave energy, then converted to electricity. Criswell has since demonstrated that LSP could supply Earth's entire power grid more cheaply than Earth-based solar power, if implemented on a large scale, and without generating chemical waste on Earth. Impressed with Criswell's idea, lunar astronaut John Young advocated strongly for LSP in his later years, and today the Japanese Shimizu Corporation is gearing up to create a PVC array, to be called Luna Ring, that will span the Moon's circumference.

Off-Earth industry and exploration may depend most on water, made of hydrogen and oxygen, useful as rocket propellant and breathing gas. Launched in 1994 to observe the Moon and near-Earth asteroid 1620 Geographos, the Clementine spacecraft was a partnership between NASA and the Strategic Defense Initiative Organization, a government group whose main purpose was to develop missile defense. The objective was to test advanced equipment, but in 1998 NASA announced that an improvised test utilizing Clementine's radar showed evidence of water in the form of ice in the Aiken Basin near the lunar South Pole.

SEE ALSO: Lunar Prospector and Surface Ice (1998), Preparing for New Missions (2018), Building a Lunar Infrastructure (2019–44)

A concept drawing of the Clementine probe.

LUNAR PROSPECTOR
AND SURFACE ICE

While Clementine was in space, another probe called Lunar Prospector was under development to launch as part of NASA's Discovery Program. Carrying five instruments, Prospector was to make low-altitude polar orbits around the Moon, mapping the chemical composition and gravitation of the surface below, measuring radon release from different regions, and measuring the Moon's weak magnetic field. On July 18, 1997, when the instrument suite was being prepared in the US, astrogeology pioneer Eugene Shoemaker was north of Alice Springs, Australia, studying meteorite impact craters with his wife, astronomer Carolyn Shoemaker (b. 1929). Driving around a curve on a dirt road, they were hit by another vehicle. Carolyn survived, but Gene did not.

When Prospector began its lunar mission on January 7, 1998, Shoemaker's ashes were aboard with the instrument suite. In light of Clementine's detection of possible water ice, Prospector's neutron spectrometer (NS) was of particular interest, as it could locate hydrogen, which should be detectable in any regions with hidden ice. The NS did reveal hydrogen, and thus water, a great deal of it, mixed with the regolith at both lunar poles.

Subsequent to the Prospector mission, additional space probes would add to the evidence for water ice at the lunar surface. In 2008, for instance, a NASA instrument carried on India's Chandrayaan-1 lunar orbiter would detect hydrogen above large areas of the Moon. In 2009, NASA would send a rocket stage from its LCROSS probe to impact Cabeus crater near the lunar South Pole. Spectral analysis would show the ejecting debris consisting of more than 5 percent water.

Meanwhile, Prospector completed its mission on July 31, 1999, when it was crashed deliberately into a southern crater, laying Shoemaker's remains on the world that he had once hoped to visit. On the capsule containing his ashes was the following Shakespearean quote:

And, when he shall die, take him and cut him out in little stars. And he will make the face of heaven so fine that all the world will be in love with night and pay no worship to the garish sun.
—Romeo and Juliet, Act III, scene ii

SEE ALSO: New Understanding of Craters (1948–60), Beginnings of Lunar Science (1964), Astrogeology (1964–65)

An artist's drawing of the Lunar Prospector orbiting the Moon.

Gravity gradient
(Eötvös)

30

0

-30

NEW GENERATION
OF MOON PROBES

THE EARLY TWENTY-FIRST CENTURY witnessed a flowering of unpiloted, scientific lunar probes. In 2003, the European Space Agency's SMART-1 orbiter mapped chemical elements across the lunar landscape. Four years later, NASA initiated the ARTEMIS mission by redirecting two Earth-orbiting probes to orbit the Moon's Lagrangian points, locations where lunar and Earth gravitation are balanced. That same year, China, India, and Japan all launched lunar missions. In October 2007, China launched the lunar orbiter Chang'e-1 and six years later the Chang'e-3 lander, which delivered a rover to the lunar surface. Meanwhile, Japan launched the Selene orbiter, also called Kaguya, and a year later India launched Chandrayaan-1, carrying a NASA instrument that would detect hydrogen, and thus water, above large areas of the Moon. In 2009, on the same rocket that carried the LCROSS probe that would find yet more evidence for lunar water, NASA launched the Lunar Reconnaissance Orbiter (LRO). Orbiting in a polar lunar orbit for many years, LRO has taken a host of high-resolution images of structures ranging from potential sites for future industry to Apollo lunar module descent stages.

In 2009 and 2010, images from Selene-Kaguya revealed bright areas known as "skylights," openings to what are thought to be lava tubes. Instead, these lava tubes are underground, and thus offer protection against radiation, meteoroid impacts, and temperature extremes for future lunar bases. Conceivably, sections of lava tubes could be sealed. In a process known as paraterraforming, they could be pressurized and given an Earth-like biosphere, using plants and microorganisms genetically modified to thrive in lunar regolith, turning the regolith into soil, and recycling oxygen, nitrogen, carbon, and water.

Another lunar mission during this period was NASA's Gravity Recovery and Interior Laboratory (GRAIL). Launched in 2011, GRAIL consisted of two craft orbiting the Moon in tandem. Measuring the constantly changing distance between the two craft enabled the creation of an exquisitely accurate lunar-gravity map that can be helpful to future landing missions. In 2013, NASA launched the LADEE orbiter to study structure and composition of the very thin lunar atmosphere.

SEE ALSO: Extended Missions (1971), Preparing for New Missions (2018)

These maps depict gravity gradients on the farside of the Moon, with red and blue indicating areas where gravity gradients are the strongest. They were created using measurements taken from GRAIL.

PREPARING FOR NEW MISSIONS

I N 2018, PREPARATIONS WERE UNDER WAY FOR India's Chandrayaan-2, consisting of a lunar orbiter, lander, and rover; for China's Chang'e-5 lunar sample collection mission; and for NASA's first test flight of its new launch vehicle, the Space Launch System. Together with announcements by private companies of upcoming lunar ventures, these mission preparations suggested that answers lay ahead for long-standing questions about the Moon's origin. One question involves the Apollo "KREEP" samples. These have been interpreted as evidence that the Moon had a molten era, and as evidence that the hypothesized impacting planet of the giant impact hypothesis left traces of itself in the lunar crust. Future expeditions to unexplored lunar sites could reveal whether KREEPs represent a global lunar phenomenon or a cosmic red herring.

Meanwhile, the matching elemental isotope ratios between the Moon and Earth, the physics of Earth's spin and the Moon's orbit, and detection of water within the glass beads of Apollo 15 and Apollo 17 have triggered various tweakings of the giant impact hypothesis, paving the way for different lunar-origin models. One such model, the multiple small impact hypothesis, entered the scene from 2015 to 2017. In place of the Mars-sized Theia, researchers at Israel's Weizmann Institute have hypothesized a series of roughly twenty small impacts, ejecting debris that gradually coalesced into moonlets, which in turn coalesced into the Moon that we know and love.

Lunar-origin scenarios involving impacts do not explain why the same process did not provide Earth's neighboring planet Venus with a similar moon. Impact events, after all, were frequent throughout the inner Solar System, and Venus is nearly the same mass as Earth. On the other hand, Venus rotates backward with respect to its revolution around the Sun, a situation that may be the result of an impact opposite the direction of the original Venusian spin. Planetary scientists think that such an impact, as well as previous impacts, could have provided Venus with one or more moons, but that tidal forces from the backward-spinning planet subsequently pulled a Venusian moon inward for an eventual crash, just as Earth's forward rotation continuously moves our Moon away.

SEE ALSO: Formation of the Moon (4.5 Billion Years Ago), Mission to Taurus-Littrow (1972)

A computer drawing of NASA's Space Launch System, designed to bunch the crewed Orion and cargo. As of October 2018, the SLS program was reported as being behind schedule and over-budget.

BUILDING A
LUNAR INFRASTRUCTURE

ASTRONOMY FROM THE LUNAR FARSIDE without interference from terrestrial electromagnetic signals, processing of lunar ore, lunar bases, research on how low gravity affects physiology and reproduction—these projects seemed imminent when astronauts first landed on the Moon in 1969, but they had faded a quarter century later, just prior to Clementine's discovery of lunar surface ice.

NASA, the European Space Agency (ESA), and private companies have been looking to use a new generation of megarockets to launch piloted lunar flybys, lunar orbital missions, and flights to build and visit a small space station positioned in cis-lunar space (a near-Moon location useful for staging travel to the Moon or deeper into space). Meanwhile, the Russian Federal Space Agency (ROSCOMOS) has announced that it may place its own station in lunar orbit, rather than participating directly with the cislunar station project. Beginning in the 2020s, astronauts will fly in NASA's new Orion spacecraft, powered by an ESA-built service module and launched by NASA's Space Launch System (SLS). Private companies SpaceX and Blue Origin are developing their own megarockets. The largest, SpaceX's Big Falcon Rocket (BFR), will exceed the launch thrust even of the most advanced version of the SLS, but may not be ready for several years.

This is not a new "space race." As during Apollo, there is government–industry codependence, plus NASA and ESA will conduct research, while SpaceX plans to launch tourists on lunar flybys, yet depends on NASA as a client. Meanwhile, Bigelow Aerospace is developing an inflatable lunar orbit habitat that could support NASA, serve a tourism function, or both. Still another company, MoonEx, is developing lightweight robotic lunar landers for science experiments.

As for human landings, timelines are less predictable, but large landers are expected by the 2030s, when ESA plans to build "Moon Village." Housing one hundred scientists by 2040, Moon Village will require regenerative life-support systems and high-pressure, dust-protected spacesuits with advanced gloves that don't yet exist. But it may lay the foundation for a lunar colony some time around the 75th anniversary of Apollo 11.

SEE ALSO: Studying Lunar Resources (1980s–90s), Preparing for New Missions (2018)

This artist's conception of a lunar base shows a structure constructed using 3-D printing, a possibility that has been explored by the ESA, which is planning to construct a "Moon Village" by 2030.

References

4.5 Billion Years Ago: Formation of the Moon
Hartmann, W. K., and D.R. Davis. "Satellite-sized planetesimals and lunar origin." *Icarus* 24 (1975), 504–515.
Rufu, R., O. Aharonson, and H.B. Perets. "A multiple-impact origin for the Moon." *Nature Geoscience* 10 (2017), 89–94.
Tyson, P. PBS Nova Online. "Origins." https://www.pbs.org/wgbh/nova/tothemoon/origins.html.

4.5 Billion Years Ago: Moon–Earth Pulling Begins
Tyson, N.D. "The Tidal Force." Hayden Planetarium (1995). http://www.haydenplanetarium.org/tyson/read/1995/11/01/the-tidal-force.

4.3–3.7 Billion Years Ago: The Moon and the Origin of Earth Life
Dodd, M.S., D. Papineau, T. Grenne, J.F. Slack, M. Rittner, F. Pirajno, J. O'Neil, and C.T. Little. "Evidence for early life in Earth's oldest hydrothermal vent precipitates." *Nature* 543(7643) (2017):60–64. doi: 10.1038/nature21377.
Dorminey, B. "Without the Moon, Would There Be Life on Earth?" *Scientific American* (2009). https://www.scientificamerican.com/article/moon-life-tides/.
Sharp, T. "How Far is the Moon?" (2017). https://www.space.com/18145-how-far-is-the-moon.html.

4.3–3.9 Billion Years Ago: Impacts Carve into Lunar Crust
Spudis, P., et al. "Moon 101 Lecture Series" (2008). Lunar Planetary Institute. Houston, TX. https://www.lpi.usra.edu/lunar/moon101/. Accessed October 2, 2017.

3.9–3.1 Billion Years Ago: A Lunar Facelift
Spudis, P., et al. "Moon 101 Lecture Series" (2008). Lunar Planetary Institute. Houston, TX. https://www.lpi.usra.edu/lunar/moon101/. Accessed October 12, 2017.

3.8–3.5 Billion Years Ago: Peak of Lunar Volcanic Activity
Needham, D.A., and D.A. Kring. "Lunar volcanism produced a transient atmosphere around the ancient Moon." *Earth and Planetary Science Letters* 478 (15) (2017): 175–178.
Spudis, P., et al. "Moon 101 Lecture Series" (2008). Lunar Planetary Institute. Houston, TX. https://www.lpi.usra.edu/lunar/moon 101/. Accessed October 24, 2017.

3.2–1.1 Billion Years Ago: Eratosthenian Period of Lunar Geological Time
Sagan, C., A. Druyan, and S. Soter. *Cosmos: A Personal Voyage* (1980). Episode 1: "The Shores of the Cosmic Ocean."
Spudis, P., et al. "Moon 101 Lecture Series" (2008). Lunar Planetary Institute. Houston, TX. https://www.lpi.usra.edu/lunar/moon101/. Accessed November 5, 2017.

1.1 Billion Years Ago: Copernican Period of Lunar Geological Time Begins
Sagan, C., A. Druyan, and S. Soter. *Cosmos: A Personal Voyage* (1980). Episode 7: "The Backbone of Night."
Spudis, P., et al. "Moon 101 Lecture Series" (2008). Lunar Planetary Institute. Houston, TX. https://www.lpi.usra.edu/lunar/moon 101/. Accessed November 20, 2017.

450 Million Years Ago: Impact Forms the Aristarchus Crater
"NASA—Striated Blocks in Aristarchus Crater" (2011). NASA. https://www.nasa.gov/mission_pages/LRO/multimedia/lroimages/lroc-20110216-aristarchus.html.

440–1.5 Million Years Ago: Lunar Assistance for Intelligent Land Life
Australian Academy of Science. "The Goldilocks Planet: Why Earth is our oasis." https://www.science.org.au/curious/space-time/goldilocks-planet

c. 8,000 BCE: Mesolithic Lunar Calendar
Smith, R. "World's Oldest Calendar Discovered in U.K." *National Geographic* (2013).

23rd Century BCE: Humanity's First Author
Druyan, A., and S. Soter. *Cosmos: A Spacetime Odyssey* (2014). Season 1, Episode 11: "The Immortals."
Mchale-Moore, R. "The Mystery of Enheduanna's Disk." https://janes.scholasticahq.com/article/2431.pdf.

22nd Century BCE: Moon Meets the Sun Over China
Odenwald, S. "Ancient Eclipses in China." NASA Goddard Space Flight Center (2009). https://sunearthday.nasa.gov/ 2009eclipse/ancienteclipses.php.
"Solar Eclipses and Science in Early China." https://michaelsaso.org/solar-eclipses-and-science-in-early-china/.

22nd–21st Century BCE: Sumerian Lunar Calendars
"Astronomy the Babylonian Way." http://adsbit.harvard.edu/cgi-bin/nph-iarticle_query?bibcode=2012JRASC.106.108A&db_key=AST&page_ind=0&data_type=GIF&type=SCREEN_VIEW&classic=YES.

18th–17th Centuries BCE: Complex Lunar Calendar Systems
Joseph, B. "History of Cosmology in Western Civilization." University of Hawaii. http://www.ifa.hawaii.edu/users/joseph/1.%20Babylonians.pdf.

c. 900–700 BCE: Lunar Cults in the Bible
Key, A.F. "Traces of the Worship of the Moon God Sîn Among the Early Israelites." *Journal of Biblical Literature* 84:1 (March 1965), 20–26.
Pardes, I. *Countertraditions in the Bible: A Feminist Approach* (1983).

763 BCE: Assyrian Eclipse
NASA. *Eclipse History*. https://
eclipse2017.nasa.gov/eclipse-
history. Accessed August 27, 2018.

**747–734 BCE: Nabonassar
Standardizes the Lunar Calendar**
Britton, J. *Arch. Hist. Exact Sci.* 61:
83 (2007). https://doi.org/10.1007/
s00407-006-0121-9

**c. 7th Century BCE: Earliest Mention
of Selene**
The Core Curriculum. *Sappho: 630
BCE–570 BCE*. Columbia College.
https://www.college.columbia.edu/
core/content/sappho. Accessed
June 4, 2018.
D'Aulaire, I., and E.P. D'Aulaire.
*Ingri and Edgar Parin d'Aulaire's
Book of Greek Myths*. Garden City:
Doubleday, 1962.

**6th Century BCE: Beginnings of
Nonreligious Astronomy**
Feynman, R., *The Character of Physical
Law* (Cambridge: M.I.T. Press),
46–47. Out of print; available at
University of Virginia http://people.
virginia.edu/~ecd3m/1110/Fall2014/
The_Character_of_Physical_Law.
pdf. Accessed August 30, 2018.
Sagan, C., A. Druyan, and S. Soter.
Cosmos: A Personal Voyage (1980).
Episode 7: "The Backbone of Night."
TehPhysicalist. "Feynman: 'Greek'
versus 'Babylonian' mathematics."
YouTube video, 10:19, Posted May
2012. https://www.youtube.com/
watch?v=YaUlqXRPMmY.

6th Century BCE: Thales Stops a War
COSMOS—*The SAO Encyclopedia of
Astronomy*: Thales http://astronomy.
swin.edu.au/cosmos/T/Thales.
Accessed April 30, 2018.
Sagan, C., A. Druyan, and S. Soter.
Cosmos: A Personal Voyage (1980).
Episode 7: "The Backbone of Night."

6th Century BCE: Spherical Harmony
Sagan, C., A. Druyan, and S. Soter.
Cosmos: A Personal Voyage (1980).
Episode 7: "The Backbone of Night."

**5th Century BCE: Anaxagoras Stands
Trial**
Stanford Encyclopedia of Philosophy:
Anaxagoras. https://plato.stanford.
edu/entries/anaxagoras/. Accessed
April 30, 2018.

**5th Century BCE: Greeks Understand
Lunar Phases**
American Physical Society News. "This
Month in the History of Physics"
(2006). https://www.aps.org/
publications/apsnews/200606/
history.cfm. Accessed May 1, 2018.
Graham, D.W. "Advances in Early
Greek Astronomy." http://citeseerx.
ist.psu.edu/viewdoc/download?doi=
10.1.1.573.7893&rep=rep1&type=pdf.
Accessed September 1, 2018.
"The Moon." *The Galileo Project*.
Rice University. http://galileo.rice.
edu/sci/observations/moon.html.
Accessed August 15, 2018.

**c. 350 BCE: Earth's Curved Shadow
on the Moon**
American Physical Society News. "This
Month in the History of Physics"
(2006). https://www.aps.org/
publications/apsnews/200606/
history.cfm. Accessed May 1, 2018.
Sagan, C., A. Druyan, and S. Soter.
Cosmos: A Personal Voyage (1980).
Episode 7: "The Backbone of Night."

**c. 350 BCE: Heavenly Perfection
Corrupted**
Sagan, C., A. Druyan, and S. Soter.
Cosmos: A Personal Voyage (1980).
Episode 7: "The Backbone of Night."
"The Moon." *The Galileo Project*.
Rice University. http://galileo.rice.
edu/sci/observations/moon.html.
Accessed August 15, 2018.

**Early 3rd Century BCE: The Library
of Alexandria**
"Raising Alexandria." *Smithsonian
Magazine*. https://www.
smithsonianmag.com/science-
nature/raising-alexandria-
151005550/. Accessed April 25, 2018.

**3rd Century BCE: Aristarchus
Measures Lunar Diameter and
Distance**
"Bucknell University Astronomy 101
Specials: Aristarchus and the Size of
the Moon." https://www.eg.bucknell.
edu/physics/astronomy/astr101/
specials/aristarchus.html. Accessed
December 10, 2017.

**3rd Century BCE: Quarter-Phase
Moon and Heliocentrism**
Cornell University. *Aristarchus*.
http://astrosun2.astro.cornell.
edu/academics/courses/astro201/
aristarchus.htm. Accessed
December 10, 2017.
Phys.Org. "What is the heliocentric
model of the universe?" https://
phys.org/news/2016-01-
heliocentric-universe.html.
Accessed December 20, 2018.

**3rd Century BCE: Eratosthenes
Calculates Earth's Circumference**
American Physical Society News. "This
Month in the History of Physics"
(2006). https://www.aps.org/
publications/apsnews/200606/
history.cfm. Accessed May 1, 2018.
Sagan, C., A. Druyan, and S. Soter.
Cosmos: A Personal Voyage (1980).
Episode 1: "The Shores of the
Cosmic Ocean."

3rd Century BCE: *The Sand Reckoner*
Acknowledgement: Gratitude to
*Dr. Richard Carrier, historian, for his
insight.*
Archimedes. *The Sand Reckoner*.
Translation. Department of
Mathematics, Trent University.
http://euclid.trentu.ca/math/
sb/3810H/Fall-2009/The-Sand-
Reckoner.pdf.
Phys.Org. "What is the heliocentric
model of the universe?" https://
phys.org/news/2016-01-
heliocentric-universe.html.
Accessed December 20, 2018.

**2nd Century BCE: Applying Math to
the Lunar Orbit**
Toomer, G.J. "Hipparchus on the
Distances of the Sun and Moon."
JSTOR Archive for History of
Exact Sciences, 14:2 (31.XII.1974),
126–142. https://www.jstor.org/
stable/41133426?seq=1#page_scan_
tab_contents

**c. 100 BCE: The Antikythera
Mechanism**
Edmunds, M.G., and P. Morgan. "The
Antikythera Mechanism: Still a
mystery of Greek astronomy?"
Astronomy & Geophysics, 41:6,
1 December 2000, 6.10–6.17.
https://doi.org/10.1046/j.1468-
4004.2000.41610.x.

1st–2nd Century CE: On the Face in the Moon's Orb

Plutarch. *Concerning the Face Which Appears in the Orb of the Moon.* Translation. University of Chicago. http://penelope.uchicago.edu/Thayer/E/Roman/Texts/Plutarch/Moralia/The_Face_in_the_Moon*/A.html. Accessed January 3, 2018.

Spudis, P.D. (2014). *Air and Space Magazine.* "Apollo 15 and the Power of Inspiration." https://www.airspacemag.com/daily-planet/apollo-15-and-power-inspiration-180952095/. Accessed January 3, 2018.

c. 150: *The Almagest*

Swetz, F.J. *Mathematical Treasure: Ptolemy's Almagest.* Mathematical Association of America. https://www.maa.org/press/periodicals/convergence/mathematical-treasure-ptolemy-s-almagest. Accessed January 3, 2018.

500–800: Eastern Astronomers Keep Looking Up

Al-Khalili, J. *Pathfinders: The Golden Age of Arabic Science.* Penguin, 2010.

Billard, R. "Aryabhata and Indian astronomy." http://adsabs.harvard.edu/abs/1977InJHS..12..207B. Accessed January 4, 2018

9th–11th Centuries: Shukuk

Al-Khalili, J. *Pathfinders: The Golden Age of Arabic Science.* Penguin, 2010.

11th Century: Seeing the First Sliver of a New Moon

Al-Khalili, J. *Pathfinders: The Golden Age of Arabic Science.* Penguin, 2010.

13th Century: A New Model for Lunar Motion

Al-Khalili, J. *Pathfinders: The Golden Age of Arabic Science.* Penguin, 2010.

14th Century: Lunar Brightness to Estimate Stellar Distances

Ne'eman, Y. "Astronomy in Sefarad." Tel Aviv University. http://wise-obs.tau.ac.il/judaism/sefarad.html. Accessed February 1, 2018.

14th Century: Adjusting Lunar-Distance Variation

Al-Khalili, J. *Pathfinders: The Golden Age of Arabic Science.* Penguin, 2010.

1543: The Moon Orbits Alone

Al-Khalili, J. *Pathfinders: The Golden Age of Arabic Science.* Penguin, 2010.

"Copernican System." The Galileo Project. Rice University. http://galileo.rice.edu/sci/theories/copernican_system.html. Accessed January 15, 2018.

1570s: Moon and Sun Orbit Earth

"The Astronomers: Tycho Brahe and Johannes Kepler. Ice Core Records—From Volcanoes to Supernovas." Harvard University. http://chandra.harvard.edu/edu/formal/icecore/The_Astronomers_Tycho_Brahe_and_Johannes_Kepler.pdf. Accessed December 28 2017.

1581: A Dream of a Lunar Voyage

"The Astronomers Tycho Brahe and Johannes Kepler. Ice Core Records—From Volcanoes to Supernovas." Harvard University. http://chandra.harvard.edu/edu/formal/icecore/The_Astronomers_Tycho_Brahe_and_Johannes_Kepler.pdf. Accessed December 28, 2017.

1609: Telescopic Study of the Moon Begins

Rice University. "The Galileo Project: Thomas Harriot." http://galileo.rice.edu/sci/harriot.html. Accessed January 3, 2018.

Rice University. "The Galileo Project: Galileo." http://galileo.rice.edu/galileo.html. Accessed January 3, 2018.

17th Century: Advancing Telescopes Eye the Moon More Closely

"The First Telescopes." American Institute of Physics. https://history.aip.org/exhibits/cosmology/tools/tools-first-telescopes.htm. Accessed February 1, 2018.

Late 17th Century: The Moon Inspires Isaac Newton

Isaac Newton Institute for Mathematical Sciences: The Isaac Newton Institute. https://www.newton.ac.uk/about/isaac-newton. Accessed February 5, 2018.

18th Century: Improving Instruments Advance Lunar Astronomy

Wilson, C. (2003). "Astronomy and Cosmology." In R. Porter (ed.), *The Cambridge History of Science*, pp. 328–353. Cambridge: Cambridge University Press. doi:10.1017/CHOL9780521572439.015.

Rice University. "The Galileo Project: Edmund Halley." http://galileo.rice.edu/Catalog/NewFiles/halley.html. Accessed February 6, 2018.

Late 18th Century: A Lunar Society in Birmingham

The Lunar Society. "Historic UK." https://www.historic-uk.com/CultureUK/The-Lunar-Society/. Accessed January 10, 2018.

"Lunar Society of Birmingham and their circle." https://www.npg.org.uk/collections/search/group/1188. Accessed January 10, 2018.

1824: Another Doctor Turns His Eyes to the Moon

"On the Moon with James Nasmyth, 1874" (2014). http://web.mit.edu/redingtn/www/netadv/SP20141020.html. Accessed December 5, 2018.

"The Man Who Found a City on the Moon" (1990). *The Aurora.* SAO/NASA Astrophysics Data System.

1870s: Verne Inspires the Father of Astronautics

"Konstantin E. Tsiolkovsky." National Aeronautics and Space Administration. https://www.nasa.gov/audience/foreducators/rocketry/home/konstantin-tsiolkovsky.html. Accessed January 12, 2018.

Redd, N.T. (2013). "Konstantin Tsiolkovsky: Russian Father of Rocketry." Space.com. https://www.space.com/19994-konstantin-tsiolkovsky.html. Accessed January 12, 2018.

Siddiqi, A. (2007). "Russia's Long Love Affair with Space." *Air and Space.* https://www.airspacemag.com/space/russias-long-love-affair-with-space-19739095/. Accessed December 15, 2017.

1873–1909: Scientists Consider Lunar Origins

Brush, S.G. "Early History of Selenogony." In *Origin of the Moon*. Hartmann, W.K., R.G. Phillips, and G.J. Taylor, eds. Houston: Lunar and Planetary Institute, 1986, 3–15.

Brush, S.G. "Nickel for Your Thoughts: Urey and the Origin of the Moon." *Science* 217 (1982): 891–898.

1914–22: The Moon Proves General Relativity

Overbye, D. "The Eclipse That Revealed the Universe." *New York Times*, July 31, 2017. https://www.nytimes.com/2017/07/31/science/eclipse-einstein-general-relativity.html. Accessed March 1, 2018.

1926: First Liquid-Fueled Rocket

"Dr. Robert H. Goddard, American Rocketry Pioneer." National Aeronautics and Space Administration. https://www.nasa.gov/centers/goddard/about/history/dr_goddard.html. Accessed February 18, 2018.

Izdatel'stvo Akademii Nauk. "Theory of Space Flight" (Teoriya kosmicheskogo poleta) SSSR. Leningrad, 1932. Translated (Jerusalem, 1971) by Israel Program for Scientific Translations as "Interplanetary Flight and Communication" (Mezhplanetnye soobshcheniya) by Rynin NA. Published Pursuant to an Agreement NASA. https://ntrs.nasa.gov/archive/nasa/casi.ntrs.nasa.gov/19720015160.pdf. Accessed December 15, 2017.

Neufeld, M. (2016). "Robert Goddard and the First Liquid-Propellant Rocket." *Smithsonian Air and Space*. https://airandspace.si.edu/stories/editorial/robert-goddard-and-first-liquid-propellant-rocket. Accessed February 17, 2018.

Redd, N.T. (2013). "Robert Goddard: American Father of Rocketry." https://www.space.com/19944-robert-goddard.html. Accessed December 1, 2017.

"Robert Hutchings Goddard." Biographical Note. Clark University Archives. https://www2.clarku.edu/research/archives/goddard/bio_note.cfm. Accessed February 17, 2018.

1929: *Woman in the Moon*

Izdatel'stvo Akademii Nauk. "Theory of Space Flight" (Teoriya kosmicheskogo poleta) SSSR. Leningrad, 1932. Translated (Jerusalem, 1971) by Israel Program for Scientific Translations as "Interplanetary Flight and Communication" (Mezhplanetnye soobshcheniya) by Rynin NA. Published Pursuant to an Agreement NASA. https://ntrs.nasa.gov/archive/nasa/casi.ntrs.nasa.gov/19720015160.pdf. Accessed December 15, 2017.

Moore, K. (2014). "Reviews: Frau im Mond" (1929). *Starburst*. https://www.starburstmagazine.com/reviews/frau-im-mond-1929. Accessed January 4, 2018.

National Aeronautics and Space Administration: "Hermann Oberth." https://www.nasa.gov/audience/foreducators/rocketry/home/hermann-oberth.html. Accessed January 3, 2018.

Siddiqi, A. (2007). "Russia's Long Love Affair with Space." *Air and Space*. https://www.airspacemag.com/space/russias-long-love-affair-with-space-19739095/. Accessed December 15, 2017.

1938: BIS Lunar Spaceship Design

"The BIS Lunar Spaceship." The British Interplanetary Society. https://www.bis-space.com/what-we-do/projects/bis-lunar-spaceship. Accessed January 6, 2018.

"Lindbergh's Anti-Jewish Speech Meets with Severe Criticism in American Press." September 1941. Jewish Telegraphic Agency. https://www.jta.org/1941/09/15/archive/lindberghs-anti-jewish-speech-meets-with-severe-criticism-in-american-press. Accessed December 16, 2017.

1930–44: Origins of the Saturn V Moon Rocket

"A Brief History of Rocketry." NASA Kennedy Space Center. https://science.ksc.nasa.gov/history/rocket-history.txt. Accessed December 15, 2017.

Mallon, T. (2007). "Rocket Man: The complex orbits of Wernher von Braun." *New Yorker*. https://www.newyorker.com/magazine/2007/10/22/rocket-man. Accessed December 1, 2017.

Neufeld, M. *Von Braun: Dreamer of Space, Engineer of War*. Vintage, 2008.

1945: Operation Overcast

Adams, G., and D. Balfour. *Unmasking Administrative Evil*. Sage Publications, 1998.

Lower, Wendy. "Willkommen" (book review). *New York Times*. https://www.nytimes.com/2014/03/02/books/review/operation-paperclip-by-annie-jacobsen.html. Accessed January 10, 2018.

Mallon, T. (2007). "Rocket Man: The complex orbits of Wernher von Braun." *New Yorker*. https://www.newyorker.com/magazine/2007/10/22/rocket-man. Accessed December 1, 2017.

"Peenemunde–1943." Global Security.org. https://www.globalsecurity.org/wmd/ops/peenemunde.htm. Accessed September 26, 2018.

Thornton, M. "Rocket Expert Renounces U.S. In Nazi Probe." *Washington Post*. https://www.washingtonpost.com/archive/politics/1984/10/18/rocket-expert-renounces-us-in-nazi-probe/c4ba2ea9-4b47-4489-8bc1-3f7963c3205f/?noredirect=on&utm_term=.3919e7b09fb2. Accessed January 14, 2014. This is about Arthur L.H. Rudolph, a close friend and co-worker of Wernher Von Braun.

"World War II: Operation Paperclip." Jewish Virtual Library. https://www.jewishvirtuallibrary.org/operation-paperclip. Accessed January 6, 2018.

1948–60: New Understanding of Craters

"Eugene M. Shoemaker." National Academy of Sciences. http://www.nasonline.org/publications/biographical-memoirs/memoir-

pdfs/shoemaker-eugene.pdf.
Accessed January 18, 2018.

1957: Sputnik
Cadbury, D. *Space Race*.
    HarperCollins, 2005.
Chaikin, A. "History of NASA
    Missions: A Scientific, Engineering
    & Human Adventure." Presentation
    at NASA Goddard Space Flight
    Center, June 2017. http://www.
    andrewchaikin.com/wp-content/
    uploads/downloads/2017/06/NASA-
    Missions-Human-1-of-3-web.pdf.
    Accessed March 3, 2018.
Russian Space Web. "Sergei Korolev."
    http://www.russianspaceweb.com/
    korolev.html. Accessed February
    1, 2018.
Sagdeev, R. "Sputnik and the Soviets."
    Science 318 (5847, 2007), 51–52.
    DOI: 10.1126/science.1149240.

1958: Explorer 1
Cadbury, D. *Space Race*.
    HarperCollins, 2005.
NASA. "Explorer 1 Overview."
    http://mail.cosmicevolution.
    net/?_task=mail&_mbox=INBOX.
    Accessed February 2, 2018.

**1958–59: A New Discovery and a
    New Agency**
NASA. "The Birth of NASA." https://
    www.nasa.gov/exploration/
    whyweexplore/Why_We_29.html.
    Accessed January 16, 2018.
"Studying the Van Allen Belts 60 Years
    After America's First Spacecraft."
    NASA. https://www.nasa.gov/
    feature/goddard/2018/studying-
    the-van-allen-belts-60-years-after-
    america-s-first-spacecraft. Accessed
    March 2, 2018.

**1959: First Pictures of the Moon's
    Farside**
Cadbury, D. *Space Race*. HarperCollins
    2005.
Zarya. "The Mission of Luna 3." http://
    www.zarya.info/Diaries/Luna/
    Luna03.php. Accessed February 15,
    2018.

1961: Humans Enter Space
Cadbury, D. *Space Race*.
    HarperCollins, 2005.

Hanser, K. "Mercury Primate Capsule
    and Ham the Astrochimp" (2015).
    Smithsonian National Air and Space
    Museum. https://airandspace.
    si.edu/stories/editorial/mercury-
    primate-capsule-and-ham-
    astrochimp. Accessed February 1,
    2018.
Redd, N.T. (2012). "Yuri Gagarin:
    First Man in Space|The Greatest
    Moments in Flight." Space.com.
    https://www.space.com/16159-
    first-man-in-space.html. Accessed
    February 1, 2018.

1961: An American in Space
Cadbury, D. *Space Race*.
    HarperCollins, 2005.
Chaikin, A. "History of NASA
    Missions: A Scientific, Engineering
    & Human Adventure. Presentation
    at NASA Goddard Space Flight
    Center" (June 2017). http://www.
    andrewchaikin.com/wp-content/
    uploads/downloads/2017/06/NASA-
    Missions-Human-1-of-3-web.pdf.
    Accessed March 3, 2018.

1962: Planning Lunar Missions
"Enchanted Rendezvous: The
    Lunar-Orbit Rendezvous
    Concept." SP-4308 SPACEFLIGHT
    REVOLUTION. NASA History
    Office. https://history.nasa.gov/SP-
    4308/ch8.htm. Accessed February
    12, 2018.

**1962: The Moon Speech at Rice
    Stadium**
John F. Kennedy Moon Speech—Rice
    Stadium. NASA. https://er.jsc.nasa.
    gov/seh/ricetalk.htm. Accessed
    August 1, 2017.

1963: Human Computers
Berman, E. "Why the Soviets Beat the
    U.S. in Sending a Woman to Space"
    (2015). http://time.com/3891625/
    first-woman-space/. Accessed
    December 2, 2018.
Neufeld, M. "Katherine Johnson,
    Hidden Figures, and John Glenn's
    Flight." *Smithsonian Air and Space*.
    https://airandspace.si.edu/stories/
    editorial/glenn-johnson-hidden-
    figures. Accessed December 2, 2017.

**1963–64: Saturn Architecture Takes
    Shape**
"SP-4206: Stages to Saturn." NASA
    History Office. https://history.nasa.
    gov/SP-4206/ch1.htm. Accessed
    March 1, 2018.

**1964: Two's Company, Three's a
    Crowd**
Cadbury, D. *Space Race*. New York:
    HarperCollins, 2005.
Chaikin, A. "History of NASA
    Missions: A Scientific, Engineering
    & Human Adventure." Presentation
    at NASA Goddard Space Flight
    Center, June 2017. http://www.
    andrewchaikin.com/wp-content/
    uploads/downloads/2017/06/NASA-
    Missions-Human-1-of-3-web.pdf.
    Accessed March 3, 2018.
Zak, A. "World's first space crew
    flies riskiest mission ever!"
    Russian Space Web. http://www.
    russianspaceweb.com/voskhod.
    html. Accessed March 7, 2018.

1964: Beginnings of Lunar Science
Granath, B. (2014). "Pioneer 4 Marked
    NASA's First Exploration Mission
    Beyond Earth." NASA. https://
    www.nasa.gov/content/pioneer-
    4-marked-nasas-first-exploration-
    mission-beyond-earth. Accessed
    March 12, 2018.
Hall, R.C. "A History of Project
    Ranger." NASA History Office, 1977.
    https://history.nasa.gov/SP-4210/
    pages/Info.htm#I_Top. Accessed
    March 9, 2018.
"Luna Mission." Lunar and Planetary
    Institute. https://www.lpi.usra.edu/
    lunar/missions/luna/. Accessed
    March 12, 2018.
Zak, A. "Planetary Moon Missions."
    Russian Space Web. http://www.
    russianspaceweb.com/spacecraft_
    planetary_lunar.html. Accessed
    March 3, 2018.

1964–65: Astrogeology
"Eugene M. Shoemaker." National
    Academy of Sciences. http://
    www.nasonline.org/publications/
    biographical-memoirs/memoir-
    pdfs/shoemaker-eugene.pdf.
    Accessed January 18, 2018.
Phinney, W.C. (2015). "Science

Training History of the Apollo Astronauts." Lunar and Planetary Institute. https://www.lpi.usra.edu/lunar/strategies/Phinney_NASA-SP-2015-626.pdf. Accessed January 5, 2018.

"Scientists in the Astronaut Corps." NASA History Office. https://www.hq.nasa.gov/pao/History/SP-4214/ch5-10.html. Accessed February 27, 2018.

1965: Improving Space Capabilities

Cadbury, D. *Space Race*. New York: HarperCollins, 2005.

Leonov, A. (2005). "Learning to Spacewalk: A cosmonaut remembers the exhilaration—and terror—of his first space mission." https://www.airspacemag.com/space/the-nightmare-of-voskhod-2-8655378/. Accessed December 17, 2017.

"Timeline of Earth Spaceflights." European Space Agency. http://www.esa.int/About_Us/Welcome_to_ESA/ESA_history/50_years_of_humans_in_space/Timeline_of_early_spaceflights. Accessed September 17, 2018.

"What Was the Gemini Program?" NASA, 2011. https://www.nasa.gov/audience/forstudents/5-8/features/nasa-knows/what-was-gemini-program-58.html. Accessed January 14, 2018.

1965–66: Learning to Rendezvous and Dock

Aldrin, B. *Line-of-sight guidance techniques for manned orbital rendezvous*. Sc.D. thesis, Massachusetts Institute of Technology, Dept. of Aeronautics and Astronautics, 1963.

Granath, B. "Gemini's First Docking Turns to Wild Ride in Orbit" (2016). NASA. https://www.nasa.gov/feature/geminis-first-docking-turns-to-wild-ride-in-orbit. Accessed February 6, 2018.

"What Was the Gemini Program?" NASA, 2011. https://www.nasa.gov/audience/forstudents/5-8/features/nasa-knows/what-was-gemini-program-58.html. Accessed January 14, 2018.

1966: Neutral Buoyancy

Neufeld, M.J., and J.B. Charles. "Practicing for space underwater: inventing neutral buoyancy training, 1963–1968." *Endeavour* 39:3–4. Online July 15, 2015. https://airandspace.si.edu/files/pdf/research/neufeld-charles-neutral-buoyancy.pdf. Accessed December 20, 2017.

"What Was the Gemini Program?" NASA, 2011. https://www.nasa.gov/audience/forstudents/5-8/features/nasa-knows/what-was-gemini-program-58.html. Accessed January 14, 2018.

1966: Tragedies

Cadbury, D. *Space Race*. New York: HarperCollins, 2005.

"Luna Mission." Lunar and Planetary Institute. https://www.lpi.usra.edu/lunar/missions/luna/. Accessed March 12, 2018.

"Lunar Orbiter 1." NASA Solar System Exploration. https://solarsystem.nasa.gov/missions/lunar-orbiter-1/in-depth/. Accessed January 7, 2018.

"Remembering NASA Astronauts Elliot See and Charles Bassett." NASA, 2016. https://www.nasa.gov/feature/remembering-nasa-astronauts-elliot-see-and-charles-bassett. Accessed December 1, 2017.

"Sergei Korolev." Russian Space Web. http://www.russianspaceweb.com/korolev.html. Accessed February 1, 2018.

"What Was the Gemini Program?" NASA, 2011. https://www.nasa.gov/audience/forstudents/5-8/features/nasa-knows/what-was-gemini-program-58.html. Accessed January 14, 2018.

1967: Apollo 1 Fire

Chaikin, A. "Apollo's Worst Day." *Air and Space Magazine*, 2016. https://www.airspacemag.com/history-of-flight/apollo-fire-50-years-180960972/. Accessed Septermber 12, 2017.

Chaikin, A. "History of NASA Missions: A Scientific, Engineering & Human Adventure." Presentation at NASA Goddard Space Flight Center, June 2017. http://www.andrewchaikin.com/wp-content/

uploads/downloads/2017/06/NASA-Missions-Human-1-of-3-web.pdf. Accessed March 3, 2018.

Larimer, S. "'We have a fire in the cockpit!' The Apollo 1 disaster 50 years later." *Washington Post*, January 26, 2017. https://www.washingtonpost.com/news/speaking-of-science/wp/2017/01/26/50-years-ago-three-astronauts-died-in-the-apollo-1-fire/?utm_term=.f74bdacf8411. Accessed September 12, 2017.

1967: Reengineering Apollo

Chaikin, A. "History of NASA Missions: A Scientific, Engineering & Human Adventure." Presentation at NASA Goddard Space Flight Center, June 2017. http://www.andrewchaikin.com/wp-content/uploads/downloads/2017/06/NASA-Missions-Human-1-of-3-web.pdf. Accessed March 3, 2018.

"NASA Apollo Mission Apollo-1—Investigation—NASA History Office." https://history.nasa.gov/Apollo204/inv.html. Accessed August 15, 2017.

1967: Declaring Peace on the Moon

"Status of International Agreements Relating to Activities in Outer Space." United Nations Office of Outer Space Affairs. http://www.unoosa.org/oosa/en/ourwork/spacelaw/treaties/status/index.html. Accessed July 26, 2017.

"Treaty on Principles Governing the Activities of States in the Exploration and Use of Outer Space, including the Moon and Other Celestial Bodies." United Nations Office of Outer Space Affairs. http://www.unoosa.org/oosa/en/ourwork/spacelaw/treaties/introouterspacetreaty.html. Accessed July 26, 2017.

1968: Lunar Tortoises

"An Evening with the Apollo 8 Astronauts" (Annual John H. Glenn Lecture Series). *Smithsonian*, November 2008. https://www.youtube.com/watch?v=Q2h_FtLzrrU. Accessed December 23, 2017.

"50 Years Ago: Solving the Pogo Effect." NASA. https://www.nasa.gov/feature/50-years-ago-solving-the-pogo-effect.

"50 Years Ago, Zond-5: A prototype of the Soviet crew capsule loops behind the Moon!" Russian Space Web. http://www.russianspaceweb.com/zond5.html. Accessed September 21, 2018.

Zond Mission. Lunar and Planetary Science Institute. https://www.lpi.usra.edu/lunar/missions/zond/. Accessed January 2, 2018.

1968: Reaching for the Moon

"Apollo 7 (AS–205): First manned test flight of the CSM." Smithsonian National Air and Space Museum. https://airandspace.si.edu/explore-and-learn/topics/apollo/apollo-program/orbital-missions/apollo7.cfm. Accessed November 28, 2017.

"Apollo 8: Christmas on the Moon." NASA, 2014. https://www.nasa.gov/topics/history/features/apollo_8.html. Accessed December 14, 2017.

"50 Years Ago: Solving the Pogo Effect." NASA. https://www.nasa.gov/feature/50-years-ago-solving-the-pogo-effect. Accessed July 21, 2018.

Jones, E.M., and K. Glover (eds.). "A Visit with the Snowman." Apollo 12 Lunar Surface Journal. https://www.hq.nasa.gov/alsj/a12/a12.summary.html. Accessed December 13, 2018.

1968: Earthrise

"An Evening with the Apollo 8 Astronauts" (Annual John H. Glenn Lecture Series). Smithsonian, November 2008. https://www.youtube.com/watch?v=Q2h_FtLzrrU. Accessed December 23, 2017.

"Apollo 8: Christmas on the Moon." NASA, 2014. https://www.nasa.gov/topics/history/features/apollo_8.html. Accessed December 14, 2017.

1969: Dress Rehearsals

"Apollo 10." NASA, 2009. https://www.nasa.gov/mission_pages/apollo/missions/apollo10.html. Accessed January 17, 2018.

Howell, E. (2018). "Apollo 9: The Lunar Module Flies." https://www.space.com/17616-apollo-9.html. Accessed September 21, 2018.

1969: One Giant Leap

"July 20, 1969: One Giant Leap for Mankind." NASA History Office. https://www.nasa.gov/mission_pages/apollo/apollo11.html. Accessed November 25, 2017.

Spudis, P., et al. (2008). "Moon 101 Lecture Series." Lunar Planetary Institute. Houston, TX. https://www.lpi.usra.edu/lunar/moon101/. Accessed November 23, 2017.

Teitel, A.S. (2016). "This Rocket Failed to Put Soviets on the Moon." *Popular Science*. https://www.popsci.com/this-rocket-failed-to-put-soviets-on-moon. Accessed December 12, 2017.

1969: Beginnings of Lunar Field Science

"Apollo Lunar Landing and Sample Return." NASA Planetary Protection. https://planetaryprotection.nasa.gov/missions-past/apollo/. Accessed December 2, 2017.

"Apollo Lunar Surface Experiments Package, ALSEP Familiarization Course Handout for Training." NASA, 1969. https://www.hq.nasa.gov/alsj/ALSEP-1969FamHandout.pdf. Accessed January 7, 2018.

"Apollo 12 Mission: Surveyor III Analysis. Lunar and Planetary Institute." https://www.lpi.usra.edu/lunar/missions/apollo/apollo_12/experiments/surveyor/. Accessed December 15, 2015.

Rummel, J.D., J.H. Allton, and D. Morrison. "A Microbe on the Moon? Surveyor III and Lessons Learned for Future Sample Return Missions." Lunar and Planetary Science Conference presentation, 2011. https://www.lpi.usra.edu/meetings/sssr2011/pdf/5023.pdf. Accessed January 28, 2018.

Spudis, P., et al. (2008). Moon 101 Lecture Series. Lunar Planetary Institute. Houston, TX. https://www.lpi.usra.edu/lunar/moon 101/. Accessed December 5, 2017.

Warmflash, D., M. Larios-Sanz, J. Jones, G.E. Fox, and D.S. McKay. "Biohazard potential of putative Martian organisms during missions to Mars." Aviat Space Environ Med. 2007 Apr; 78(4 Suppl):A79–88.

1969: Making Moonquakes

"Apollo Lunar Sample Analysis." Lunar and Planetary Institute. https://www.lpi.usra.edu/lunar/samples/. Accessed December 2017.

Spudis, P., et al. (2008). Moon 101 Lecture Series. Lunar Planetary Institute. Houston, TX. https://www.lpi.usra.edu/lunar/moon101/. Accessed December 17, 2017.

Yamada, R. Description of Apollo Seismic Experiments. Japan Aerospace Exploration Agency. https://darts.isas.jaxa.jp/planet/seismology/apollo/The_Description_of_Apollo_Seismic_Experiments.pdf.

1969: The Lunar Receiving Laboratory

"The Lunar Samples." Lunar and Planetary Science Institute. https://www.lpi.usra.edu/publications/books/moonTrip/viTheLunarSamples.pdf. Accessed Nobember 30, 2017.

Mangus, S., and W. Larsen. "Lunar Receiving Laboratory Project History. NASA/CR–2004–208938." 2004. https://www.lpi.usra.edu/lunar/documents/lunarReceivingLabCr2004_208938.pdf.

Meyer, C. (2003). "Lunar Sample Mineralogy. NASA Lunar Petrographic Educational Thin Section Set." https://curator.jsc.nasa.gov/lunar/letss/mineralogy.pdf. Accessed November 29, 2017.

Taylor, G.R., R.C. Simmonds, and C.H. Walkinshaw. "SP–368 Biomedical Results of Apollo. Chapter 2: Quarantine Testing and Biocharacterization of Lunar Materials." https://history.nasa.gov/SP-368/s5ch2.htm. Accessed November 1, 2017.

1970: A Successful Failure

"Apollo 13." NASA, 2009. https://www.nasa.gov/mission_pages/apollo/missions/apollo13.html.

Cellania, M. (2010). "Forty Years Ago: Apollo 13." http://mentalfloss.com/

article/24441/forty-years-ago-apollo-13. Accessed January 13, 2018.

Kranz, G. *Failure Is Not an Option: Mission Control from Mercury to Apollo 13 and Beyond.* Simon & Schuster, 2009.

Lovell, J., and J. Kruger. *Lost Moon: The Perilous Voyage of Apollo 13.* Houghton Mifflin, 1994.

1971: Return to the Moon

"Apollo 14 Landing Site Overview." Lunar and Planetary Sciences Institute. https://www.lpi.usra.edu/lunar/missions/apollo/apollo_14/landing_site/. Accessed November 11, 2017.

Spudis, P., et al. (2008). "Moon 101 Lecture Series." Lunar Planetary Institute. Houston, TX. https://www.lpi.usra.edu/lunar/moon101/. Accessed December 20, 2017.

Yamada, R. "The Description of Apollo Seismic Experiments." Japan Aerospace Exploration Agency. https://darts.isas.jaxa.jp/planet/seismology/apollo/The_Description_of_Apollo_Seismic_Experiments.pdf.

1971: Extended Missions

Anzai, T., M.A. Frey, and A. Nogami. "Cardiac arrhythmias during long duration spaceflights." *Journal of Arrhythmia* 2014; 30(3): 139–149.

"Apollo 15 Mission Report." NASA. https://www.hq.nasa.gov/alsj/a15/ap15mr.pdf. Accessed November 28, 2017.

Teitel, A.S. "NASA's (Un)Censored Moonwalkers." *Popular Science*, 2014. https://www.popsci.com/nasas-uncensored-moonwalkers. Accessed December 2, 2014.

Williams, D.R. "Apollo 15 Hammer-Feather Drop." https://nssdc.gsfc.nasa.gov/planetary/lunar/apollo_15_feather_drop.html.

1972: Descartes Highland

"Apollo 16 Science Experiments." Lunar and Planetary Science Institute. https://www.lpi.usra.edu/lunar/missions/apollo/apollo_16/experiments/. Accessed January 12, 2018.

"Mysterious Descartes." Lunar and Planetary Science Institute. https://www.lpi.usra.edu/publications/books/rockyMoon/17Chapter16.pdf. Accessed January 20, 2018.

Spudis, P., et al. (2008). "Moon 101 Lecture Series." Lunar Planetary Institute. Houston, TX. https://www.lpi.usra.edu/lunar/moon101/. Accessed December 21, 2017.

1972: Mission to Taurus-Littrow

"Apollo 17 Mission Report." National Aeronautics and Space Administration, 1973. https://www.hq.nasa.gov/alsj/a17/A17_MissionReport.pdf. Accessed February 14, 2018.

Saal, A.E., E.H. Hauri, M.L. Cascio, J.A. Van Orman, M.C. Rutherford, and R.F. Cooper. "Volatile content of lunar volcanic glasses and the presence of water in the Moon's interior." *Nature.* 2008;454(7201):192–195. doi: 10.1038/nature07047.

Spudis, P., et al. (2008). "Moon 101 Lecture Series." Lunar Planetary Institute. Houston, TX. https://www.lpi.usra.edu/lunar/moon101/. Accessed December 21, 2017.

STEM Talk. Episode 4: Harrison Schmitt discusses being the first scientist on the moon. https://www.youtube.com/watch?v=IM7UVCHraHs. Accessed December 19, 2017.

1972: Apollo Biostack

Acknowledgement: Michael D. Delp, Ph.D., of Florida State University: Gratitude for providing insights during a phone conversation with the author.

Delp, M.D., J.M. Charvat, C.L. Limoli, R.K. Globus, and P. Ghosh. "Apollo Lunar Astronauts Show Higher Cardiovascular Disease Mortality: Possible Deep Space Radiation Effects on the Vascular Endothelium." *Sci Rep.* 2016; 6: 29901.

Graul, E.H., et al. "Radiobiological results of the Biostack experiment on board Apollo 16 and 17." *Life Sci Space Res.* 1975; 13: 153–159.

"The Deadly Van Allen Belts?" NASA. https://www.nasa.gov/sites/default/files/files/SMIII_Problem7.pdf. Accessed March 25, 2018.

1972–74: Cancelled Apollo Missions

"Wednesday's Child." Apollo Applications. NASA History Office. https://history.nasa.gov/SP-4208/ch3.htm. Accessed March 30, 2018.

"Apollo 18 through 20–The Cancelled Missions." NASA. https://nssdc.gsfc.nasa.gov/planetary/lunar/apollo_18_20.html. Accessed March 31, 2018.

"Manned Venus Flyby." NASA, 1967. https://ntrs.nasa.gov/archive/nasa/casi.ntrs.nasa.gov/19790072165.pdf. Accessed March 31, 2018.

"Scientific rationale summaries for Apollo candidate lunar exploration landing sites–NASA Report. NASA, 1970. https://ntrs.nasa.gov/archive/nasa/casi.ntrs.nasa.gov/19790073898.pdf. Accessed March 31, 2018.

Spudis, P., et al. (2008). "Moon 101 Lecture Series." Lunar Planetary Institute. Houston, TX. https://www.lpi.usra.edu/lunar/moon101/. Accessed February 4, 2018.

1970s–80s: Elucidating Lunar History

"Future Chinese Lunar Missions." NASA Goddard Spaceflight Center. https://nssdc.gsfc.nasa.gov/planetary/lunar/cnsa_moon_future.html. Accessed September 26, 2018.

Hartmann, W.K., and D.R. Davis. "Satellite-sized planetesimals and lunar origin 1975." *Icarus* 24, 504–515.

"Luna Mission." Lunar and Planetary Institute. https://www.lpi.usra.edu/lunar/missions/luna/. Accessed March 21, 2018.

Saal, A.E., E.H. Hauri, M.L. Cascio, J.A. Van Orman, M.C. Rutherford, and R.F. Cooper. "Volatile content of lunar volcanic glasses and the presence of water in the Moon's interior." *Nature* 2008; 454(7201): 192–195. doi: 10.1038/nature07047.

Spudis, P., et al. (2008). "Moon 101 Lecture Series." Lunar Planetary Institute. Houston, TX. https://www.lpi.usra.edu/lunar/moon101/. Accessed March 30, 2018.

1980s–1990s: Studying Lunar Resources

"Clementine Project Information." NASA Goddard Spaceflight Center. https://nssdc.gsfc.nasa.gov/planetary/clementine.html. Accessed January 12, 2018.

Criswell, D.R. "Development and Commercialization of the Lunar Solar Power System." IAF abstracts, 34th COSPAR Scientific Assembly. Second World Space Congress, October 10–19, 2002, Houston, TX. Abstract 262.

Meyer, C. (2003). "Lunar Sample Mineralogy." NASA Lunar Petrographic Educational Thin Section Set. https://curator.jsc.nasa.gov/lunar/letss/mineralogy.pdf. Accessed February 20, 2017.

Papike, J., L. Taylor, and S. Simon. Chapter 5 in *Lunar Sourcebook: A user's guide to the Moon*. Lunar and Planetary Institute. https://www.lpi.usra.edu/publications/books/lunar_sourcebook/pdf/Chapter05.pdf. Accessed March 2, 2018.

Simko, T., and M. Gray. "Lunar Helium-3 Fuel for Nuclear Fusion: Technology, Economics, and Resources." *World Futures Review* 2014. https://doi.org/10.1177/1946756714536142.

1998: Lunar Prospector and Surface Ice

"Eugene Shoemaker Ashes Carried on Lunar Prospector." NASA Jet Propulsion Laboratory. https://www2.jpl.nasa.gov/sl9/news82.html. Accessed November 22, 2017.

"The Lunar Prospector Mission." Lunar and Planetary Science Institute. https://www.lpi.usra.edu/lunar/missions/prospector/. Accessed January 17, 2018.

2003–13: New Generation of Moon Probes

Haruyama, J., et al. "Detection of lunar lava tubes by lunar radar sounder onboard Selene (Kaguya)." Lunar and Planetary Science Conference presentation, 2017. https://www.hou.usra.edu/meetings/lpsc2017/pdf/1711.pdf. Accessed January 15, 2018.

"Missions to the Moon." The Planetary Society. http://www.planetary.org/explore/space-topics/space-missions/missions-to-the-moon.html. Accessed March 25, 2018.

"Moon Missions." NASA. https://moon.nasa.gov/exploration/moon-missions/. Accessed May 1, 2018.

Redd, N.T. "Chang'e-4: Visiting the Far Side of the Moon." Space.com. https://www.space.com/40715-change-4-mission.html. Accessed June 1, 2018.

2018: Preparing for New Missions

"Future Chinese Lunar Missions." NASA Goddard Spaceflight Center. https://nssdc.gsfc.nasa.gov/planetary/lunar/cnsa_moon_future.html. Accessed April 10, 2018.

Musser, G. "Double Impact May Explain Why Venus Has No Moon." *Scientific American*, 2006. https://www.scientificamerican.com/article/double-impact-may-explain/. Accessed February 12, 2018.

Rufu, R., O. Aharonson, and H.B. Perets. "A multiple-impact origin for the Moon." *Nature Geoscience* 10, 89–94 (2017).

2019–44: Building a Lunar Infrastructure

"Moon Village." European Space Agency. https://www.esa.int/About_Us/Ministerial_Council_2016/Moon_Village. Accessed May 20, 2018.

"NASA's Lunar Outpost Will Extend Human Presence in Deep Space." NASA. https://www.nasa.gov/feature/nasa-s-lunar-outpost-will-extend-human-presence-in-deep-space. Accessed May 15, 2018.

'Space Launch System." NASA. https://www.nasa.gov/exploration/systems/sls/index.html. Accessed May 1, 2018.

Index

Image Credits

AKG: 150

Alamy: Chronicle: 24; Classic Image: 32; Dinodia Photos: 70; Heritage Image Partnership, Ltd.: 52, 126; The History Collection: 44, 68; INTERFOTO: 90; Keystone Pictures USA: 156; Theodore Iiasi: 58; NC Collections: 134; North Wind Picture Archives: 62; Pictorial Press Ltd.: 118; RGB Ventures/SuperStock: 142; Ronald Grant Archive: 108; Science History Images: 40, 76

Art Resource: © Trustees of the British Museum: 34

British Interplanetary Society: 110

ESA: AOES Medialab, ESA 2002: xii; Foster + Partners: 198

Courtesy of Vince Gaffney: 20

Granger Historical Picture Archive: 38

iStock: blinow61: 72; duncan1890: 2; Stephan Hoerold: 116; Madhourse: 100; merznatalia: 4; Nastasic: 86

Robert Knudsen/John F. Kennedy Presidential Library and Museum, Boston: 132

Library of Congress: 54

Look and Learn: 94

Metropolitan Museum of Art: 26

NASA: cover, front endpapers single, iii, 6, 10, 14, 98, 106, 120, 122, 128, 130, 136, 144, 146, 148, 152, 164, 184; Ames: 192, back cover top right; GSFC: 56, 190; JPL-Caltch/CSM: 194; JSC: xix, 16, 102, 154, 158, 160, 162, 166, 168, 170, 172, 174, 176, 178, 180, 182, back endpapers spread, back cover (top left); Robert Markowitz: 48; MSFC: 112, 186

Private Collection: 80

Science Source: © David A. Hardy: 188; © Mikkel Juul Jensen: 18; NASA/JPL: 140

Shelby White and Leon Levy Archives Center/Institute for Advanced Study, Princeton, NJ, USA: 104

Shutterstock.com: Eti Ammos: 84; markara: 66

Tim Donovan: 212

Wellcome Collection: 28, 42, 46, 78, 92, back cover (bottom right)

Courtesy of Wikimedia Commons: front endpapers, 12, 36, 50, 74, 82, 88, 96, back endpapers single, back cover (bottom left), spine; Kevin Gill: 196; mefman00: 22; Mlevitt1: 30; NASA: 8, 138; RKK Energia: 124; US Army: 114; Mogi Vicentini: 64; Yale: 60

About the Author

DR. DAVID WARMFLASH is an astrobiologist, space medicine researcher, and science communicator. He received his MD from Tel Aviv University's Sackler School of Medicine, and has done post-doctoral work at Brandeis University, the University of Pennsylvania, and the NASA Johnson Space Center (JSC), where he was part of the NASA's first cohort of astrobiology postdoctoral fellows. At JSC, David worked and studied under the late Dr. David S. McKay, who helped found NASA's astrobiology program and decades earlier trained astronauts to conduct lunar field geology.

In the early 2000s, David served as a member of NASA's Jupiter Icy Moons Orbiter Science Definition Team and has collaborated with The Planetary Society on three experiments that have sent microbial samples into space. These included "the peace experiment," in which an Israeli and a Palestinian student worked as co-investigators with Dr. Warmflash and other colleagues to study microbial samples that flew on NASA's STS-107. In 2011, David collaborated with The Planetary Society on its efforts to include an astrobiology experiment on Roscosmos's Phobos-Grunt probe, which attempted to make a round-trip to the Martian moon Phobos.

Inspired by Dr. Carl Sagan from childhood into his adult years, David has written numerous articles in scientific journals and popular science publications, including *Wired UK, Scientific American, Air and Space,* and *Leaps Magazine.* His pieces cover a range of science topics, from the search for extraterrestrial life and space exploration to genetics, neuroscience, biotechnology, the origins of life, and the history of science. In 2017, David contributed a short essay about the feasibility of human hibernation to *George and the Blue Moon* by Lucy Hawking and Stephen Hawking. *Moon: An Illustrated History* is David's first book.